FLORIDA STATE
UNIVERSITY LIBRARIES

AUG 1 8 2000

TALLAHASSEE, FLORIDA

VISIONS OF SUSTAINABILITY

Visions of Sustainability
Stakeholders, change and indicators

STEPHEN MORSE
International Development Centre, University of Reading, UK
NORA McNAMARA
MOSES ACHOLO
BENJAMIN OKWOLI
Diocesan Development Services, Idah, Nigeria

Ashgate
Aldershot • Burlington USA • Singapore • Sydney

© Stephen Morse, Nora McNamara, Moses Acholo and Benjamin Okwoli 2000

All rights reserved. No part of this publication may be reproduced, stored in a retrieval system or transmitted in any form or by any means, electronic, mechanical, photocopying, recording or otherwise without the prior permission of the publisher.

Published by
Ashgate Publishing Limited
Gower House
Aldershot
Hampshire GU11 3HR
England

Ashgate Publishing Company
131 Main Street
Burlington
Vermont 05401
USA

Ashgate website: http://www.ashgate.com

British Library Cataloguing in Publication Data
Visions of sustainability : stakeholders, change and
    indicators
    1. Sustainable development - Nigeria
    I. Morse, Stephen
    333.7'09669

Library of Congress Catalog Card Number: 00-131232

ISBN 1 84014 867 5

Printed and bound by Athenaeum Press, Ltd.,
Gateshead, Tyne & Wear.

Contents

List of Figures vi
List of Tables vii
Foreword x
Acknowledgements xi

Introduction 1

1 Agricultural Sustainability: A brief tour 5

2 Rationale and Method 46

3 Eroke Village 77

4 Some Life Histories 114

5 Production and Consumption 134

6 Livelihood and Leisure 178

7 Visions of Sustainability 209

Appendix A 221
Appendix B 230
Appendix C 232

References *236*
Index *244*

List of Figures

1.1	Simplified example of an agro-ecosystem	42
1.2	Spectrum of cropping systems	43
1.3	Over-yielding in intercrops	44
1.4	Explanation of the Land Equivalent Ratio (LER) with two crops (a and b)	45
2.1	Sketch map of Nigeria showing the approximate location of the major cities and towns referred to in the text	74
2.2	Igalaland: major towns, roads (solid lines) and state boundaries (dotted line)	75
2.3	Sketch map of Imane and the 'Abo' group of villages	76
3.1	Map of Eroke village. Major roads and footpaths are shown along with the compounds	112
4.1	Sketch map of West Africa showing the relative locations of Nigeria and Equatorial Guinea	133
5.1	Proportion of land under cultivation for six Eroke farms	175
6.1	Comparison of median ranks of male and female income and expenditure	207
6.2	Eroke Farmer Councils (FC's) timeline	208

List of Tables

1.1	United Nations sustainability indicators (SIs) for sustainable agriculture	34
1.2	Some sustainability indicators proposed for Sub-Saharan African agro-ecosystems	36
1.3	Three sets of indicators relating to the sustainability of tree home gardens	37
1.4	Summary of a scheme based on the quantification of constraints to sustainability	38
1.5	Sustainability Indicators based on five aspects of cabbage production in Malaysia	40
1.6	Example of the use of Weighted Goal Programming applied to a set of twenty Sustainability Indicators (SIs) developed for a farming system in Burundi	41
2.1	A summary of some of the Nigerian based 'village/community' studies that were used to feed into aspects of the Eroke study between 1992 and 1998	68
2.2	List of key resource persons used in the Eroke study (1992 to 1998)	69
2.3	Summary of surveys used in the Eroke study (1992 to 1998)	70
3.1	Number of Madakis and Gagos in the 'Abo' group of villages	106
3.2	Traditional councils in Kogi State, Nigeria	106
3.3	Population of the 'Abo' group of villages and Eroke	107
3.4	Household population and population change for a selected sample of 20 households in Eroke	108
3.5	Some of the common crop species found in Eroke	109
3.6	Some common economic and firewood trees in Eroke	110

4.1	Some details for the thirteen families described in the life histories	132
5.1	Mean yields and standard deviations (SD) for some common crops found in Eroke	156
5.2	Some crop mixtures in Eroke	157
5.3	Calculation of the partial Land Equivalent Ratio (pLER) and Land Equivalent Ratio (LER) for the maize/cassava intercrop in Eroke	158
5.4	Perceptions of agricultural change over a 10 year period (late 1980s to 1998)	160
5.5	Crop areas for five farms over the period 1988 to 1997	162
5.6	Tree crop inheritance patterns for women in Eroke (before 1970 to 1990)	164
5.7	Average cultivation intensities (R-factors), standard deviation (SD) and coefficient of variation (CV) for six farms in Eroke	165
5.8	Perceived change in adoption of coping strategies for dealing with problems of soil fertility between the late 1980s and 1998	166
5.9	Change in length of fallow period	166
5.10	Male and female involvement in farm activities	167
5.11	Sources of farm labour for male and female farmers	167
5.12	Two sets of analyses designed to determine the relationship between labour input with land area	168
5.13	Preferences in 1998 when using labour for farm work	170
5.14	Membership of labour groups in the late 1980s compared with 1998	170
5.15	Perceived change in firewood and water consumption and availability (1988 to 1998)	171
5.16	Household (HH) size and HH consumption of foodstuff (kg/annum)	171
5.17	Predicted land requirement and actual areas cultivated in 1997 for a number of the staple crops in Eroke	172
5.18	A summary of production and consumption based indicators in terms of sustainability	174

6.1	Number of occupations and sources of income per respondent	197
6.2	Types of trader and goods traded in Eroke	198
6.3	Type of remittance sent back to Eroke by 'abroadians'	199
6.4	Skills and other agricultural inputs brought back by migrants	199
6.5	Inflation in education and health care costs (1988 to 1998)	200
6.6	General inflation in Eroke (1970 to 1980 and 1980 to 1996)	201
6.7	The number of visits in two years (1993 to 1995) to three sources of medical treatment/help	202
6.8	Some of the characteristics of the eight Eroke *ojas*	203
6.9	Comparison of credit taken as cash with credit taken as kind	204
6.10	Comparison of the end of year balances in Eroke with some neighbouring villages (1990 to 1993)	204
6.11	Perceived change in quality of life in Eroke (late 1980s to late 1990s)	205
6.12	A summary of livelihood and leisure indicators in terms of sustainability	206
7.1	Some suggested areas to explore within sustainability	220
8.1	Field crops	222
8.2	Tree crops	225
8.3	Change in animal ownership, disease incidence and consumption	226
8.4	Firewood and water resources	227
8.5	Coping strategies for dealing with problems of soil fertility	229

Foreword

The modern concept of development is almost fifty years old. In that period many new concepts have been introduced, each with a different emphasis. The Catholic Church has always been involved in human development, but since the second Vatican Council it has been called upon to be 'a new way of being Church'. Because of this social issues are stressed, and this in turn placed an emphasis on social analysis as a means of identifying the root causes of many contemporary problems. This book is timely for it addresses the current preoccupation of many of our donor partners in their quest for sustainability. Examining agricultural sustainability is even more critical as this is the main occupation of the Igala people. This book draws extensively on both qualitative and quantitative methods so it ought to reassure most development practitioners that this research is an integrating force illustrating interdependence as an important approach to development.

This is just a beginning. Given the conclusions reached at the close of this work, it is expected that other areas of concern in Igalaland such as health care and women's activities can be examined. Having young Igala's involved in the research is most encouraging as this needs to become integral to all church based activities. I congratulate all four people involved, and encourage them to delve deeply into this type of analysis regarding development intervention. Having a methodology for assessing it is a step forward, and I pray that each time it is adopted and replicated new insights will be possible. This work shows sustainability is still at the pioneering stage and I appreciate the courage this takes. It is also my wish that those who read this work will examine the controversial issue and enrich the debate with their revelations.

Rt. Rev. E. O. Obot
Bishop of Idah Diocese, Nigeria

Acknowledgements

We would like to thank all those who have helped us in this work for over six years. The whole Eroke community is hereby acknowledged and thanked for their extraordinary hospitality and welcome, and for truly allowing us to live with them for extended periods. Without access to their personal insights and benefits of their experience, this book would not have been possible. The community provided us with more information than can be encapsulated in this book, and we hope that this can eventually be documented so as to preserve their socio-cultural heritage for the knowledge and enjoyment of many generations to come. We would especially like to thank the Bishop of Idah, Rt. Rev. E. S. Obot for his immense support and encouragement.

Stephen Morse would like to thank the School of Development Studies, University of East Anglia, and the International Development Centre, University of Reading, for allowing him to spend time in helping to plan and implement this project. A special thanks to my family, especially my wife Maura, who have been very supportive throughout.

Nora McNamara would like to thank the Holy Rosary Sisters especially the regional and central leadership and the Idah community for their unflinching support. Without the help of the McNamara family, this book would not have been possible, so for their financial assistance and ongoing encouragement I offer a special word of praise and thanks.

Moses Acholo acknowledges the encouragement of his parents, Mr and Mrs Egbunu Acholo to participate in this work. He also wishes to thank Mrs Helen Ochimana, Elle Alaji and Anna Ameh for taking care of him during extended stays in Eroke, Simon Wodi for assisting him in the early days of data collection and the Okoliko families for arranging village meetings.

Benjamin Okwoli wishes to thank his parents for having encouraged him to follow in his father's footsteps and for encouraging him to experience village life as an essential part of understanding development.

Finally the secretarial staff of DDS deserve a special word of thanks for their unending patience at all times in producing endless amounts of questionnaires and other backup material.

Introduction

Each person occasionally needs to distance her/himself from work and arrive at a greater understanding of how to improve on results for the greater benefit of all concerned. This should lead to more appreciation of our fellow travelers and help us stay in rhythm with our inner life and longing. This is true of all workers, but perhaps more especially so for those engaged in the delicate task of development. The causes giving rise to the post second world war phenomenon called development are not addressed here, but it behoves us to remember that justice is one of the issues at stake. If development practitioners wish to redress the root causes of injustice the situation needs some measure of acceptance and tolerance of all committed to the task of development. If there isn't unity of purpose amongst this group, can relationships be built with those meant to benefit? The question is how can this be satisfactorily initiated? The participants in the research described here believed they were being asked questions they would have thought of themselves, and this gave them food for thought for years to come. Can we dare hope to do this with a wider audience? If the provocative can be evoked and responded to, the authors welcome the reaction.

Sustainability is one of the leading themes in the later part of the twentieth century. There are many good definitions, but the language of sustainability is value laden and open to misinterpretation. This has become the bane of scientists, planners and those charged with ensuring that 'sustainability happens'. How can you 'get' something unless you know when you have it? The idea of a concrete, measurable property called sustainability that can be applied objectively to development and agriculture, is a holy grail in current research. In large part this is reflection of the need for implementation, but has also been facilitated by rapid advances in information technology. Donors and development partners have become keen on sustainability throughout the 'project cycle' - from appraisal and evaluation to impact assessment. Much has been written about this, and the reader is entitled to question why another contribution is necessary. However, to date the bulk of this work describes what sustainability is and why it ought to be attained, but with little description or analysis of experience. The purpose of this work is to provide such experience.

The Diocesan Development Services (DDS) based in Idah diocese, Kogi State, Nigeria, aims at including internal evaluation as part of the reflection-action-reflection process and selects topics considered vital for development in this reflection process. Two previous publications, McNamara and Morse (1996) and McNamara and Morse (1998), describe in detail two aspects of this development work: on-farm research and the provision of financial services. Given the growing practical importance of sustainability and its measurement, DDS felt obliged to test its practicality in real life situations. Therefore a decision was made by DDS in the early 1990s to measure the sustainability of some of its initiatives in rural development. This was also an opportunity for training personnel in research methods and to provide a model that could be adapted for replication in other communities.

Although the decision was easily made, implementing it was more of a challenge given that there was very little practical guidance as to how it could be done. Part of the problem was with definition, but there was also the central difficulty of knowing what one should look at and how to go about it. Donors haven't spelt out clearly what they mean by sustainable, but there are two commonly applied schools of thought. Some see sustainable as shorthand for a set of 'good' practice, but for others it equates more or less to financial self-sufficiency on the part of the field partner (McNamara and Morse, 1998). For example, a project that focused on the promotion of compost was seen as sustainable and hence 'good', whereas a project involving the purchase and distribution of fertilizer was unsustainable and hence 'bad'.

In contrast, the research described here focused on the notion of sustainability as a system property rather than as a menu of 'good' and 'bad' practices. This had a number of advantages. Firstly, it allowed a more integrated and holistic approach, incorporating aspects of the society that hinge on and are influenced by agriculture. In terms of development funding, this was especially useful as a number of different donors with different mandates and approaches could be involved in projects within a single village. Secondly, a broader vision of sustainability as a property allowed full participation on the part of project beneficiaries and flexibility as to what they considered important. A narrow focus on specific practices is much more constraining and doesn't allow scope for other possibilities inherent in project implementation. Thirdly, much of the current literature on sustainability focuses on the 'property' aspect rather than practice.

In order to provide some focus it was initially decided to concentrate on the measurement of agricultural sustainability in one Igala village, and the starting point of the analysis was taken to be the onset of Structural Adjustment programmes (SAP) in Nigeria (1986/87). Agriculture was a natural choice, partly because it is the mainstay of the Igala economy, and also because of the abundance of literature. DDS had been involved for many years in agricultural initiatives in Igalaland (McNamara and Morse, 1997). It was decided to look at one village (called Eroke) largely because it seemed like a logical spatial unit and realistic in terms of the modalities necessary for such a study. Development from the mid 1980s onwards has been influenced by SAP and the end of the cold war which meant less interest in, and fewer funds for, development intervention in Nigeria. This has speeded up some aspects of development, but it was not favourable to those least able to help themselves. Therefore, the introduction of SAP provided a logical base from which to launch an analysis of sustainability.

Defining sustainability was more problematic. Social scientists believe in participative approaches and cultivate shared leadership methodologies. With this in mind, and because of the philosophy on which DDS is based, it was decided to generate a local theme for the meaning of sustainability. A number of key resource people were invited to discuss sustainability and what it meant to them. These included historians, anthropologists, geographers, education personnel and many with a background in sociology and related social science subjects. These warned about Igala not having a word into which sustainability could be translated and the likelihood of having it confused with self-reliance. Their word preference would be for 'viability' and not sustainability. They saw individuals, groups and communities as the main players in this new approach, and it was only in so far as these were central to the act that all other items could be weighed and quantified. A historian pointed to the critical moment in history when the main source of innovation shifted to human beings themselves and away from natural forces. From then on the human story has been increasingly one of choice in situations of continual change. Need and necessity are the mother of invention and discovery.

This book is about measuring sustainability within a development context, and it is hoped that it will be of interest to local grass root communities, development practitioners, academics, students and development partners. The social science literature is not as awash with the issue of measuring sustainability as is that of the economists and natural scientists. The latter two groups are used to dealing with matters that can be weighed, measured and quantified. Not so with the social scientists, or at least not to the same extent. They deal much more with the qualitative, in matters relating to change, decision making, and the very quality of life itself, especially in concerns of the on-going growth and development of each human being. Creating the circumstances for this type of formation is a prerequisite for the success of economists and natural scientists and the roles are complementary. The longer the years of experience the more evidence there is for this inter-dependence. This work aims at striking a balance between the use of highly quantifiable materials of the natural scientists and economists and the more qualitative, intuitive approach of the social scientists. The worlds of the two converge on the topic of sustainability. If a greater discourse can be established between these two worlds and their actors, then this book will have achieved its objectives.

The first chapter of this book will explore briefly some of the common themes in agricultural sustainability. Rather than give the reader a complete overview of this topic, something beyond the ability of text such as this, it will instead concentrate on the aspects associated with the findings of the study. The second chapter presents the methodology employed throughout the research. The third and fourth chapters are intended to give the reader an overview of Eroke. These show that sustainability is meaningful only in so far as it relates to people in real life situations. Even if the focus is on agriculture, they bring it to life. These two chapters also show the richness and diversity of life, and by doing so it is hoped to thus provide the reader with a socio-cultural and economic context within which agriculture can be analysed. The following two chapters present the more quantifiable results of the research subdivided into: production and consumption patterns, livelihood and finance. The final chapter draws upon the main findings of the research to answer the key question posed at the onset: can agricultural sustainability really be measured?

Finally, it has to be said that a field agency such as DDS undertaking such a study is in something of a cleft-stick, and one of the risks involved would be that donors might view such work as a dilution of development activity. DDS firmly believes a field agency concerned with development and justice has an obligation to engage in such work. However the greatest reward for this venture into sustainability would be the meeting of minds and hearts showing another path where unity, interdependence, trust and acceptance would be the hallmarks.

1 Agricultural Sustainability: A brief tour

Introduction

The notion that development should be sustainable was a central theme of the Earth Summit held in Rio in 1992, and of many others before and since. Two definitions of sustainability commonly quoted in the literature are:

> development that meets the needs of current generations without compromising the ability of future generations to meet their needs and aspirations.
>
> World Commission for Environment and Development (1987)
>
> development that improves the quality of human life while living within the carrying capacity of supporting ecosystems.
>
> International Union for the Conservation of Nature (1991)

Agriculture is a major human activity, providing income and sustenance for much of the rural and urban poor in developing countries, and for this reason it was given a prominent position in the Rio conference debates. Sustainability in terms of maintaining adequate production with minimal damage to the environment predates the Rio conference by many centuries. The modern form of sustainable agriculture largely arose as a reaction to accelerated change in agricultural practice that took place predominantly in Europe and North America in this century. Increasing farm sizes, loss of woodland and hedgerows, indiscriminate use of pesticides and inorganic fertilizer have had a detrimental impact on the environment for everyone to perceive. A counter attack was inevitable.

The rhetoric behind sustainability (development or agriculture) is appealing, but to be operational one must know when it has been attained (Goldman, 1995). Some have argued that the absence of sustainability (unsustainability) is far easier to recognise than its presence (Jodha, 1989). For example one can look for environmental damage. The problem is the means by which one should recognise sustainability - what do you look for, and when and where do you look?

This book is about agricultural sustainability and rural livelihoods, and how it can be practically measured. To do this, the current thinking in the debate must be summarised, and in particular some of the commonly employed associations with agricultural sustainability. The emphasis will be on crops rather than livestock, as this was what was important in the study area. Others have stressed the importance of crops in developing countries:

> Since crops are the main output of most agricultural systems, the dynamics of crop sustainability should be an initial focus of attention.

Goldman (1995)

The associations with agricultural sustainability that have been selected are: cropping systems, production, energy, soils and pests. These are headings under which agricultural sustainability is often discussed, and used here to provide the reader with an overview of the topic. At the same time issues of fundamental concern are raised. Following this is a brief introduction to the importance of people in agricultural sustainability, and an outline of progress towards identifying and gauging sustainability.

The magnitude of the sustainability debate and the diverse range of opinion means the list employed here is by no means complete. The topic of environmental valuation and how this relates to what some call weak sustainability as opposed to strong sustainability (Common and Perrings, 1992; Rennings and Wiggering, 1997) is only touched on. The importance of impact assessment is readily acknowledged, but not included here because of the context within which this work was done.

Agriculture and sustainability - an overview

It can be argued that the management of natural resources lies at the heart of the debate surrounding sustainability. The birth of ecology as a science has also been a major influence in the origin and history of sustainable development (Kidd, 1992). Ecology brought the notion of carrying capacity, a finite set of resources supporting a finite number of individuals, and helped spawn the concept of Maximum Sustainable Yield (MSY) – replacement must equal removal otherwise the population will decline. Agriculture has been particularly influential for the following reasons (Lele, 1991):

1) the end product of agriculture is often food. It is therefore one of the foundations of human society.

2) agricultural systems occupy large areas of land - far more land than any other industry with the possible exception of forestry. In the UK, for example, agriculture accounts for approximately 77 per cent of the total land area (Pearce, 1993).

3) in some countries agriculture has also undergone substantial change over the past century, moving from subsistence to what is aptly termed agri-business.

The rapid change towards agri-business has inevitably been at some cost. The desire to produce more, increase quality and decrease cost has resulted in a series of adaptations - fewer farms, larger fields and a replacement of human energy with that derived from fossil fuels. The scale and rapidity of the change meant negative effects on the environment (human and physical) were unavoidable. The result has been an increasing desire to look at alternative approaches within agriculture that may alleviate some of these negatives. Terms such as agroecology, alternative agriculture, ecological food production (Begon, 1990), low input sustainable agriculture (LISA; Beets, 1990), organic agriculture and integrated crop production have become commonplace although often confused. To each of their proponents they offer some distinctive feature, but the common denominator is widespread use of the adjective sustainable. As a counterpoint, conventional (= intensive or high input) agriculture is seen as unsustainable. There is a general notion that sustainable agriculture encompasses environmental/ecological (also referred to as biophysical), social (or human community) and economic spheres (Flora, 1992; Olson, 1992; Spencer and Swift, 1992; Yunlong and Smit, 1994). Sustainability is often seen as the classic holistic vision of nature and the world, while intensive systems are seen as simplistic, narrow in focus, with little intrinsic concern for the environment.

However, beyond this commonality, they represent a very diverse family. To take a specific example, LISA is assumed to be sustainable agriculture with an accepted low-level of artificial inputs (Reijntjes *et al.*, 1992), while 'organic agriculture' is demarcated by the absence of defined substances (mostly pesticides and artificial fertilizer) during production. Some proponents of organic agriculture may have little, if any, empathy with LISA, and may not see the use of 'sustainable' applied to that approach as truly representative. Indeed the use of the word sustainable applied to agriculture immediately throws into sharp relief the fundamental problem of definition. Much depends upon who is defining in each individual context, and as a result there are almost as many definitions as there are people who take an interest. Definitions can take very different tangents, and the following provide a taste of this diversity. More examples can be found in Dunlap *et al.* (1992), Swift and Woomer (1993), Yunlong and Smit (1994), Goldman (1995) and Hansen (1996).

> What is sustainable agriculture after all? The only sustainable agriculture is profitable agriculture. Short and sweet.
>
> Ainsworth (1989)
>
> A sustainable agriculture is one that equitably balances concerns of environmental soundness, economic viability, and social justice among all sectors of society.
>
> Allen *et al.* (1991)
>
> Sustainable agriculture refers to the use of agricultural land in such a way to ensure that over time no net quantitative or qualitative loss of natural resources occurs.
>
> Fresco and Kroonenberg (1992)

Sustainable agriculture consists of agricultural processes, that is, processes involving biological activities of growth or reproduction intended to produce crops, which do not undermine our future capacity to successfully practice agriculture.

Lehman *et al.* (1993)

These four examples all emphasise different yet often expressed aspects of the sustainability discourse. The first is centred on profitability: 'short and sweet'! The second stresses the need for fair distribution of resources and justice within society – often referred to as equitability. The third and fourth have a time dimension, and encapsulate the need to maintain natural resources and 'capacity to successfully practice'. The acknowledgement of a time dimension is often referred to as futurity. Given this diverse emphasis on profitability, equitability and futurity it may appear that a definitive answer to the question 'what is sustainable agriculture' seems as far off as ever, and the following challenge made as recently as 1994 is revealing:

Sustainable agriculture can be viewed from ecological, social and economic perspectives, and should be addressed relative to all three. Sustainable agriculture may also be viewed as a series of interacting systems at various spatial scales. Conceptual diversity exists in thinking about, and practising sustainable agriculture. Each theme implies a different emphasis and a distinct set of analytical questions. Resolving incompatibilities among these approaches, and among the goals of sustainable agriculture generally, remains a conceptual and analytical challenge.

Yunlong and Smit (1994)

There is much interest in sustainability. The bottom-line of 'farming without damage to future generations' has widespread acceptance including those who argue that conventional, high yield/high input farming is sustainable:

high yield farming is more sustainable than organic farming. The best new high-yield farming systems have much lower levels of soil erosion, the key long-term constraint on farm production since ancient times. We also have strong evidence that high-yield farming can continue producing higher and higher yields on into the future.

Avery (1995)

While a universally accepted definition may be some way off, there are two fundamental and broad visions of agricultural sustainability (Hansen, 1996; Smith and McDonald, 1998). These are:

1) sustainability as an approach
2) sustainability as a property

The former is a philosophy that drives practice. Some practices are necessary for sustainable agriculture while others are unsustainable. For example, one may hold the view that pesticides should not be used in agriculture and farming without them is more sustainable. In this case sustainable is equated with a package of good practices rather than a fundamental property of the agricultural system. The underlying assumption is that the sustainable approach has less negative impact on the environment. There is an understanding that the good practices have to be maintained, and therefore this vision has an implied, but perhaps not overtly stated, time dimension.

Sustainability as an approach is relatively easy to define and measure. Good practices are encouraged and bad practices are avoided. The definition of good and bad is subjective, and varies between individuals, but one can monitor progress by noting the implementation of such practices by farmers. Soil analysis and testing of agricultural produce for the presence of prescribed substances such as pesticide, also have a role to play. This vision of sustainability is most common in practice, and examples can be found in Grieg-Smith *et al.* (1992) and Penfold *et al.* (1995). A typical prescription for agricultural sustainability includes the following (Goldman, 1995):

- crop rotation
- crop and livestock diversification
- nutrient recycling

- natural pest control (ie. Integrated Pest Management; IPM)
- soil conservation
- low (reduced) use of fertilizer, pesticide, fossil fuel and irrigation

The second vision is more complex. It regards sustainability as a property of the agricultural system rather than a simple package of preferred practices, although practice is an important 'driving force' influencing components within the system (Figure 1.1). Broadly, this concept seeks to define the ability of the system to exist in some preferred state and continue to deliver its products over a time scale (Spencer and Swift, 1992).

> As a property of agriculture, sustainability is interpreted as either the ability to satisfy a diverse set of goals or an ability to continue through time.
>
> Smith and McDonald (1998)

Sustainability as a property is a more sophisticated vision rooted in the complex subject of systems analysis as outlined by Conway (1986), Clayton and Radcliffe (1997) and others. Components and flows (transfer functions) within the system need to be identified, and then equated to sustainability. Given the complex nature of agro-ecosystems this is not an easy task, but the vision has the appearance of being more objective than sustainability as an approach.

> Ecosystem-level concepts are the core of sustainable agriculture - both in definition and measurement.
>
> Neher (1992)

Sustainability as a system property resonates more deeply with scientists (natural and social) and through them to policy makers. It has received a great deal of attention, albeit in theoretical terms rather than practice.

It can be argued that these two visions are really two ways of looking at the same thing. This can be illustrated by asking what has to be looked for in sustainability or indeed in unsustainability? If sustainability is seen as an approach with a set of approved practices a check for their presence or absence will identify sustainability. For example, does the farmer use pesticide, and if so how much? Similarly if sustainability is seen as a property then one can identify features that equate to 'system quality' (Bell and Morse, 1998) and measure them. For example, an absence of pesticide residue from a lake contributes to good system quality, and this, in turn, will relate to farmers use of pesticide in the catchment area. Both may be associated indicators of sustainability where one has an effect on the other. Therefore the practices seen as important to the approach of sustainable agriculture impact on sustainability seen as a property.

The two visions also share common windows on sustainability, and some of these will be explored in the following sections. The first will look at cropping systems, and in particular how they differ in terms of diversity and resilience to disruption. The latter two properties are closely entwined with sustainability. Following this is a discussion of production and economic viability, and how these have been influential in efforts to gauge sustainability. The next three sections look at some of the important elements that underpin production – energy, soil and pests.

Cropping systems, resilience and sustainability

Throughout the world there are many cropping systems, and probably the simplest classification is provided as Figure 1.2. The systems are classified as single or multiple depending upon how many crops are grown on the same area of land in a growing season. 'Single' indicates one crop is grown per season, while 'multiple' means more than one crop grown. Systems at the bottom of Figure 1.2 tend to be prevalent in developed countries, while those towards the top are most common in developing countries. In reality a farmer may adopt a number of systems on a single piece of land over time, and the farm itself may be planted to a variety of systems. Exact figures for the extent of multiple cropping are difficult to obtain, due to lack of data and confusion over terminology, but some estimates are as high as 90 per cent of total crop area for Africa. Multiple cropping systems are not confined to developing countries. In the USA, double-cropping of winter wheat followed by summer soybeans as part of a zero tillage (no cultivation of the land between crops) system has been popular (over two million ha in 1986). There is also archaeological evidence of intercropping in Britain.

The complexity (expressed as biodiversity for example), of the system increases from the bottom of the spectrum to the top. Biodiversity is typically related to what is termed system resilience (ability to withstand an environmental, economic or social shock). The general perception is that systems at the top of the spectrum are most sustainable (= resilient) and those at the bottom are the most unsustainable (less resilient). Systems towards the top of Figure 1.2 often involve an intensification of resource use. A sequential cropping system may generate more yield/year than a single crop, but the extra yield is paid for by an increase in the use of resources, although one of the key resources, light, may be wasted if only a single crop was grown. There could also be a greater usage of more limited soil resources such as nutrients and water.

Research has indicated that intercrops, a type of multiple crop, give higher yields than would be expected from equivalent areas of sole crops (Innis, 1997). An illustration is provided as Figure 1.3. The reasons for this synergy are varied (Trenbath, 1974; Francis, 1989; Ofori and Stern, 1987; Innis, 1997). For example, some experiments have indicated a reduced incidence of pests, diseases and weeds in intercrops relative to sole crops. This may be due to the different crops acting as barriers to pests and diseases, or perhaps a better accumulation of natural enemies of the pest in the intercrop relative to a sole crop. The latter may be linked to the presence of alternative prey species or the attraction of more diverse plant architecture. Intercrops may be more resilient to environmental disruption as the death of one crop will free resources for other crops to utilise. The ability of other crops to compensate will depend upon their innate aggressiveness along with the time left for growth. This resilience is often seen as central to sustainability (Goldman, 1995).

> Sustainability can be defined as the ability of a system to maintain its productivity when subject to stress or perturbation.
>
> Conway (1986)

Without the use of inputs an intercrop may be more resilient (= sustainable) than a sole crop. However, given that sole crop systems are maintained and protected with inputs of pesticide, fertilizer, irrigation etc. they are also resilient and hence can be regarded as sustainable using the 'maintenance of productivity' definition of Conway (1986). It should also be remembered that farmers and their families may not be solely reliant on agriculture for a livelihood, and livelihood diversity can also increase resilience (Goldman, 1995).

Production and economic viability

The maintenance of production over time is often seen as a key feature of agricultural sustainability.

> A cropping system is sustainable if it has an acceptable level of production of harvestable yield which shows a non-declining trend from cropping cycle to cropping cycle over the long term.
>
> Izac and Swift (1994)

The aim of the agricultural process is to produce, and hence the emphasis on maintenance of production. Although it can be argued that economic viability rather than production is the critical aspect, and sustainability should equate to maintenance or improvement in economic viability (Lehman *et al.*, 1993). All the diverse cropping systems in Figure 1.2 ultimately seek to maintain production but through different means. Those at the foot of the figure utilise chemical inputs while those at the top employ more biological based approaches.

Sustainability and production have an intricate relationship, and three different but related aspects of this are:

1) the desire for quantification (measurement in numerical terms) of production
2) the maximisation of production (or more precisely, profit)
3) efficiency of resource use (nutrients, light, water) in production.

The measurement of agricultural production from a farm is relatively easy in more simplified systems – those generally deemed to be less sustainable. In multiple cropping systems, particularly those based on intercropping, measuring production can be more complex. A logical and simple method for measuring intercrop production is to sum the yields of the crops. For example in an intercrop of two species (a and b):

intercrop yield $= Y(ab) + Y(ba) = Ya + Yb$

where $Y(ab)$ = yield of crop a in the intercrop
$Y(ba)$ = yield of crop b in the intercrop

Even with a system of only two crops this approach leads to over-simplification. How can allowances be made for qualitative differences between the two crops? For example, the nutritional quality or market value of 'a' and 'b' may be different. This has been recognised, and alternative approaches in the literature involve a conversion of crop yield to energy values, nutritional or market value.

Faced with the complexity of intercrop systems, the approach of western agricultural science has primarily been to compare it with the predominant 'western' system of sole cropping. To achieve this, the Land Equivalent Ratio (LER; Figure 1.4) is used (Innis, 1997). LER is relatively easy to calculate and interpret once the yields are known. If LER = 1 then the intercrop yield can be obtained by planting the components as sole crops (i.e. no advantage to using an intercrop). For example, in Figure 1.4 where LER = 1 there is no advantage to intercropping, and the intercrop area (1 ha) can simply be divided into two half hectare plots of each sole crop and the total output would be the same. If LER > 1 then the intercrop yield is greater than that obtained by planting the components as sole crops (i.e. use of an intercrop is advantageous). If LER < 1 then the intercrop yield is less than that which can be obtained by planting the components as sole crops (i.e. the use of an intercrop is disadvantageous). LER values in the range 1.1 to 1.5 are typical for cereal/legume intercrops.

Understanding production is often seen as an essential pre-requisite for measuring sustainability, although the measurement of production in a single season is not enough in itself. What time scale to consider is of import and will be returned to later in this chapter.

Besides issues of measurement, production has another interesting link with sustainability. Simplified systems, particularly with high levels of artificial inputs, tend to be the most productive in terms of gross output. The key emphasis in such systems from a farmer's perspective may be economic viability as opposed to production. Economic viability is revenue (derived from the production of the unit) less costs. However, farmers and their families may have a diverse range of income, only part of which may be derived from farm production. This is what some term the profitability sense of economic viability (Lehman *et al.*, 1993). Lehman *et al.* (1993) also suggest that economic viability could have a self-sufficiency sense that may not necessarily equate to profitability – a sustainable farm may be economically viable in the sense of self-sufficiency although not very profitable. It may be said that the perceived emphasis on profit above all else has fathered the notion of sustainable agriculture.

Production comes at a cost of resources, and the supply of these resources needs to be maintained for production to continue. Given this linkage, some have attempted to present production and sustainability in terms of efficiency of resource use. An example is the Total Factor Productivity (TFP) for tropical farming systems (Lynam and Herdt, 1989; Spencer and Swift, 1992). The TFP is defined as:

$$TFP = \frac{\text{value of outputs from farming system}}{\text{value of inputs into farming system}}$$

Lynam and Herdt (1989) suggest that changes in TFP over time equates to a measure of sustainability (i.e. change in productive capacity of the system). In its simplest form, an increase in the TFP equates to sustainability, while a decrease indicates a decline in the resource base and hence unsustainability. Tisdell (1996) suggests a slight modification of this idea to focus instead on profitability (P) of the system:

$$P = \frac{\text{value of output - value of input}}{\text{value of input}}$$

This is fundamentally an economic approach based on the productivity of the farming system, and works as long as inputs (including the natural resource base) and outputs can be given a monetary value. Other environmental and social effects that many consider central to sustainability are not included. For example, farming can have effects on the wider environment (so called externalities; Flora, 1992) that one can cost in terms of the investment required to cancel out those effects (Pearce, 1993).

Energy and sustainability

To keep the agro-ecosystem viable there must be energy input. The energy source can be either human or animal. In the case of high input systems the energy is from products manufactured with fossil fuels. For example, it has been estimated that 36 per cent to 55 per cent of the 1998 yields of barley, wheat, potatoes and sugar beet in the UK were due to the application of fertilizer. This century has seen a substantial substitution of human/animal energy for chemical energy in the more developed countries (Stanhill, 1984), and the so-called high input systems typically occupy the bottom of Figure 1.2. Fossil fuels and mineral deposits are resources which are not renewable, and therefore high input systems which rely heavily on energy from these sources are often labelled 'unsustainable'. In addition, they represent concentrated energy and their application may result in some leakage to the environment and disruption of the existing ecosystem (locally or on a wider scale).

Sole cropping with machinery and full inputs can be very efficient and resilient in terms of productivity from a farmer's point of view. However, the efficiency of energy use (the energy ratio) is often lower relative to less intensive systems. Efficiency of energy use (energy obtained for each unit of input) is typically measured as the ratio between energy output and energy input:

$$\text{energy ratio} = \frac{\text{total energy output}}{\text{total energy input}} = \text{energy obtained for each unit of input}$$

There are many studies of energy efficiency in agriculture, and a range of examples can be found in Pimental *et al.* (1983), Briggs and Courtney (1989), Beets (1990), Tivy (1990) and Netting (1993). Slesser (1984) suggests that a relationship between energy input and energy output over a wide range of agricultural systems (mostly developed countries) can be derived. The law of diminishing returns comes into play. As more energy is put into the system there eventually comes a point when there is no net gain. For example, five times as much energy is required to produce one unit of energy as rice in the USA compared to Borneo, suggesting that the latter is far more energy efficient, although absolute production of rice in the USA is much greater than in Borneo (Tivy, 1990). Similarly, in Papua New Guinea the net farm output was 10 MJ/person/day while in the USA it was 1,500 MJ/person/day (150 times greater; Beets, 1990).

Soils and sustainability

Given the importance of soil in agriculture, it is not surprising that the literature emphasises the need for its maintenance or improvement as central to any consideration of sustainability (Hendrix *et al*, 1992; Swift and Woomer, 1993). These include the need to maintain/improve organic matter (Swift and Woomer, 1993), pH, water holding capacity and nutrient status (Hendrix *et al.*, 1992), and prevention of erosion. Striking images of agricultural failure and soil erosion have been captured, for example in Steinbeck's '*Grapes of Wrath*'. Many soil textbooks have pictures of gully erosion, and for many this represents a classic epitome of unsustainability.

Much is known about soil, and its properties such as organic matter, nutrient status, pH etc. and how these can be measured with standardised procedures. Soil organic matter content is a component of soil quality that is particularly common in the literature on agricultural sustainability (Stocking, 1994) as it has a major influence on soil structure (the ability of the soil to retain water) and soil fertility (Swift and Woomer, 1993). Organic matter has an interesting relationship with agricultural sustainability in that intensive systems sought to replace the role of organic matter with inputs such as inorganic fertilizer and irrigation. In the less intensive systems organic matter in the form of crop residue, manure or compost was added and incorporated as a matter of course. Organic matter, however, is dilute (often less than 2 per cent nutrient content on a weight for weight basis), varies as a source of plant nutrients, and its application labour intensive. Inorganic fertilizer, by contrast, is a more concentrated form of nutrients with a known and standardised content. Although this makes crop nutrition more precise and targeted, some wastage is almost inevitable and pollution can result – particularly with soluble forms such as nitrates. Given the above, the form of plant nutrients supplied (organic or inorganic) is seen by many as a key consideration in sustainability.

Rather than measuring organic matter, pH etc. the farmer can provide views regarding change in soil quality over time. Farmers have indigenous knowledge of their soils and environment, and it is also possible to observe how farmers manage their soil. Examining what farmers do may be easier than measuring soil properties, but the underlying assumption is that the presence/absence of certain practices equates to sustainability. The use of inorganic fertilizer has been mentioned, but there are many other management techniques that could be examined; presence (or absence) of fallow periods, rotations, cover crops, mulching and erosion controls to name but a few (Stocking, 1998). The first of these can be included in some notion of cultivation intensity (Stocking, 1994) by calculating the ratio:

$$\text{cultivation intensity (R - factor)} = \frac{\text{years under cultivation}}{\text{years under cultivation + fallow}} \times 100$$

The nearer this value is to 100 per cent then the less fallow there is in the system, but acceptable 'R' values depend on a range of factors such as the soil type, crops grown and levels of inputs added. For example, in savanna areas with a cropping period of between 120 and 269 days, and low levels of input, 15 per cent is suggested as a minimum value for R on the relatively low quality ferraisol/acrisol type of soil (Stocking, 1994). This corresponds to a low cultivation intensity of three years under cultivation for every twenty.

Pest control and sustainability

Pests (insects, pathogens, vertebrates etc.) are a problem as old as agriculture, although it has been argued that the pest pressure farmers face world-wide is now greater than before (Brader, 1988). Crop protection is now big business, and multinational companies compete for sales. This began with the advent of human-made organic pesticides in the 1940s. One of the best known is DDT, though it is now banned in many countries. DDT was developed prior to the Second World War and used extensively in controlling typhus-bearing lice. The DDT group (the organochlorines) has two shared characteristics. First they are fat soluble and very stable (i.e.

difficult to chemically break-down). This means that once ingested they accumulate in animal bodies. Secondly they are toxic to a wide range of insects (i.e. broad spectrum of activity), including those that are beneficial (pollinators and natural enemies of the pests that the insecticides are meant to kill). These characteristics and the large-scale and indiscriminate use of the organochlorines post World War Two inevitably lead to environmental problems. These problems were identified and predicted by scientists early on in the pesticide era, but it took Rachel Carson's book, *Silent Spring* (Carson, 1962) to bring this to the wider attention of the public.

The problems arising from the indiscriminate use of pesticides have been among the most influential in the origin of the sustainability concept (Allen and Rajotte, 1990), although others now argue that sustainability has moved away from issues surrounding crop protection to those centred on the soil (Goldman, 1996). Problems with pesticide use changed the approach of scientists (and others) to pest control, and the result by the 1960s was an emphasis on pest management as opposed to control. This embodied the notion of tolerance of a certain level of pests rather than their complete elimination. In practice, this largely meant a reduction (or even elimination) of pesticides and replacement with other methods of reducing the pest population. The resultant philosophy, called Integrated Pest Management (IPM) has had an enormous influence on the development of sustainable agriculture, and the similarity between their goals has often been alluded to.

> It is clear that IPM and sustainable agriculture belong together. Even taken separately their goals are remarkably similar.
>
> Frans (1993)

IPM has a curious, and often contradictory, relationship with the use of pesticides. It has been argued that the use of pesticides gave birth to the IPM philosophy, and the conditions where IPM has done best are where excessive use of pesticides led to many problems. IPM has been successful when unsustainability has been obvious. Some see the absence of pesticide as the soul of IPM while others see pesticides as an important tool in the IPM 'bag of tricks'. This diversity is reflected in the wider debates on agricultural sustainability. Goldman (1996) defines a category of crops 'whose sustainability is contingent on intensive pest or disease management, including the use of pesticides'. However, like IPM, the broad view is that the less pesticide used the more sustainable the system. Jansen *et al.* (1995) suggest the calculation of a biocide index that measures the total amount of pesticide applied while allowing for differences in toxicity:

$$\text{Biocide index} = \frac{\Sigma \text{ 'toxic load' for pesticide 1}}{\text{duration of land use system}}$$

Where toxic load is a function of the amount of pesticide applied each year, its concentration of active ingredient (a.i.; the toxic component), the toxicity of the a.i. and how long the a.i. remains active in the environment. The 'Σ' symbol indicates a need to calculate each individual pesticide (1, 2, 3 etc.) applied in the agro-ecosystem.

The biocide index is essentially a measure of toxic load into the system on an annual basis, and the assumption is the lower this value, the better. The problem is in the definition of an acceptable level for sustainability. Jansen *et al.* (1995) point out that 'no clear-cut relation exists between the calculated index of biocide use and the sustainability of the system, making calculation of a limit to the biocide index impossible'. To many the biocide index is irrelevant, as the amount of pesticide applied must equal zero for the system to equate to sustainability. Only if there is some acceptable degree of pesticide use is the biocide index of any relevance, but what should this level be? Is it adequate to simply aim for a reduction (50 per cent maybe) over a given period of time?

IPM can be seen as a reactive philosophy; without the problems associated with pesticide use, it is doubtful whether IPM would ever have come into being. The term IPM is open to a wide range of interpretation and some have even suggested that IPM is more easily defined in terms of what it isn't (i.e. liberal use of pesticides) than what it is (Kremer, 1994). A reactive origin, lacking in clear consensus of meaning and a tendency to define in terms of what it isn't rather than what it is are strikingly common to both IPM and sustainable agriculture.

People, development and agricultural sustainability

As agro-ecosystems are created and managed by farmers, sustainability must incorporate their aspirations. Taking a wider perspective, the aspirations of the non-farming community must also be considered. It may be argued that pressure from this group has led to the current popularity of sustainable agriculture, and the calls for protection of the environment.

Livelihood is obviously central to human life. In countries where the majority of the population are farmers, pressures for sustainability are different. In Nigeria, it is estimated that 65 per cent to 70 per cent of the population are farmers and issues of livelihood and sustainability are intertwined. Yet the majority of Nigerian farmers are not intensive users of agro-inputs and machinery, and consequently the damage to the natural vegetation has not been as apparent as in other countries. The cropping systems in Nigeria are typically those towards the top of Figure 1.2, and relatively efficient in terms of energy use. It is reasonably argued that most African farmers are already practitioners going by the prescriptive list for sustainable agriculture given earlier (Goldman, 1995). However, it is here that we see some of the most vigorous attempts to promote sustainable agriculture as part of development. There are good reasons for this. First, most of the world's biodiversity is in the tropics and a repeat of what happened in Europe could be catastrophic. Secondly, the shift towards agribusiness has social repercussions as some of the population may lose land, their source of livelihood. There are however dangers as Goldman (1995) points out:

> Diagnoses and prescriptions for sustainability imported from Westerm industrial contexts can result in serious misconception of the needs of African agricultural systems. Overuse of modern agricultural technology cannot be considered a pre-eminent threat at present or in the near future in most of sub-Saharan Africa.

Little attention is typically given to social factors within discourses on the practice of sustainable agriculture, and one only reads about its environmental, technical and economic aspects. This is surprising as the term sustainability is value-laden and far from being so objective that everyone can agree on meaning. It would appear to be logical to begin with an analysis of people's conditions and aspirations and use them to understand practice.

There is no shortage of village and community studies in Africa, including many specific to Nigeria. Examples are Berry (1993), Guyer (1981, 1992, 1996, 1997), Netting (1993), Netting and Stone (1996), Netting *et al.* (1989), Stone (1997, 1998) and Stone *et al.* (1990, 1995). Change, diversity and resilience, often in an agricultural setting are central elements to these.

Identifying and measuring sustainability

The earlier sections have looked at some broad features of agriculture that often act as windows on sustainability. The complexity of the cropping system, including its resilience, energy flows and efficiency, soil quality and pest control are just a few of these. From the foregone literature, there are clearly incongruities. An emphasis on maintenance of production as central to sustainability begs the question as to how maintenance is to be achieved. This is not mere rhetoric. If sustainable agriculture is so vital as a practical goal, some idea of what it is, together with some means of knowing when it has been achieved are necessary.

As an approach, sustainable agriculture comprises a set of good (sustainable), and bad (unsustainable) practices. A number of farmers have switched systems, as witnessed by the growth of the organic farming movement and its low-input relatives. In this vision of sustainability, research has essentially consisted of the development of a combination of good practice for a particular context.

The implementation of sustainable agriculture, when viewed as a system property, is more complex. The properties of the system and their means of measurement are important. This approach has led to the development of the now popular Sustainability Indicators (SIs). These allow the features already discussed to be measured (Smith, 1998). For example, it is possible to measure the amount of pesticide used in any given area (eg. the Biocide Index of Jansen *et al.*, 1995) and use it as SI. Low values for the Biocide Index equates to sustainability.

Most SIs relate to indicators of sustainable development in the broadest sense and not only to agriculture (Smith, 1998). SIs are grouped in various ways depending upon what dimension of sustainability is being measured.

The simplest division is into two groups:

1) **STATE** SIs. These describe the state of a variable. For example, one may determine the physical and chemical properties of soil and the pesticide concentration of water in a lake (Swift and Woomer, 1993; Jansen et al., 1995; Miller and Wali, 1995; Penfold et al., 1995).

2) **DRIVING FORCE** SIs (also referred to as pressure, process or control SIs). These measure a process or pattern that in turn influence a state SI. The biocide index is an example of a control SI.

State and control SIs can be related, but State SIs alone do not necessarily provide information as to the causes of change (Harrington, 1992a and 1992b). For example, the concentration of pesticide in drinking water may be influenced by a number of factors, and information on this is required before the actual concentration can be explained. The rate of pesticide application in any given area (control SI) will have an influence on the measured concentration of pesticide in drinking water (state SI).

Following the Rio Earth Summit, the United Nations Commission on Sustainable Development developed approximately 130 SIs for sustainable and agricultural development. These are closely linked to the forty Chapters of the Summit document, and as well as the two types listed above there is a third category:

3) **RESPONSE** SIs. These gauge policy options and other responses to change.

Response SIs focus on what governments and others do as a response to the first two indicators. An example of the UN SIs is provided in Chapter forteen of the Agenda 21 document entitled *'Promoting sustainable agriculture'*. Seven SIs have been developed for this chapter: four driving force, two state and one response (Table 1.1). The driving force SIs are as expected. Pesticide and fertilizer use as well as energy efficiency are logical choices, and the desired direction in which they should move is also relatively clear. This is not true for some of the other proposed SIs. A high proportion of land under irrigation may be seen as a proxy indicator for intensification, and some would associate this with unsustainability. However, an increase in irrigated land may be desirable. In the case of the two state SIs, the area affected by salinization (a bad system property) is clear in terms of desired direction, but what of the area of arable land per inhabitant? Does sustainability equate to more or less arable land/inhabitant? Other specific problems with these SIs are identified by the UN and listed in Table 1.1. Some are relatively minor, and solutions have already been discussed. For example, instead of estimating pesticide use for an area, the Biocide index or similar formula may be calculated which allows for differences in relative toxicity and persistence in the environment. Other problems are more complex, such as energy use in agriculture. While much research has been done on this, more is needed before a reliable data set is established.

The UN example helps illustrate some of the problems with the assessment of sustainability. Each SI presented in Table 1.1 is not easy to measure accurately. There is no social dimension, except an oblique understanding that applying the SIs will improve people's lot. Integration is another problem. What happens if some of the SIs point in a good direction while others do not? If all have to be correct then this may not matter, but can poor performance with some be compensated by progress with other? In other words, can SIs be combined to give an aggregate index of sustainability (Swift and Woomer, 1993)? It is possible to weight the individual SIs but inevitably this will be subjective and open to dispute.

There are more fundamental problems. Before a list of SIs can be created, it is necessary to determine:

1) Where are the boundaries?
2) What time scale is needed to gauge sustainability?

The UN list is intended for application by governments, but the spatial context is important. Measuring the amount of pesticide applied in one part of a country may not be related to its concentration in the local drinking water. The latter may be more influenced by application of pesticide in a watershed some miles away – perhaps even in a different country. The futurity nature of sustainablility implies an inter-generational scale, but how long is this? When should measurement begin, or in other words what is the reference point? These are key questions, but problematic as anything can be proved with an adroit selection of starting point and time scale.

A few more examples will be covered to provide a range of approaches and their pitfalls. To achieve this, the examples will be divided into two groups. The first comprises theoretical discussions of SI development for agriculture, combined with little practical application. The second consists of case studies where the SIs are applied to real systems.

Group 1. Theoretical SI frameworks

Example 1. Sub-Saharan Africa Izac and Swift (1994) developed a matrix of SIs for small-scale agriculture in sub-Saharan Africa, with the aim of providing a framework for research in this region. They attempt to address many of the central issues in sustainability, including appropriate spatial and time scales, as well as the diverse socio-economic and environment elements. The authors acknowledge the complexity of the debate surrounding suitable spatial scales, and identify the 'village-catchment' as the 'required scale of analysis of sustainability' as it is the:

> ultimate target for assessment of sustainable agriculture as this is the scale at which the individual farmer's goals and those of the community are integrated.

They define this unit as:

> an agroecosystem managed at the social scale of the village community and at the ecological scale of the small catchment.

Within the village-catchment, they define two other important scales in developing SIs: the cropping system and the farm. With regard to time scale the authors acknowledge that:

> the choice of the period of time over which the sustainability of an agroecosystem is defined and evaluated is by necessity arbitrary.

However, they go on to say that:

> the appropriate time period for assessing trends in sustainability should be defined as at least one, and preferably several, decades.

This sets ten years as the minimum time scale for measuring sustainability. Although, the selection of the village and a minimum of ten years as the spatial and temporal scales respectively can be debated, at least they have been clearly defined.

Izac and Swift identify production, or more precisely the outputs of agriculture, as central to sustainability. This emphasis, they acknowledge, is common:

> our working definition of sustainable crop production is in agreement with the usual conventions of agricultural research in being targeted on harvestable yield as a measure of agricultural performance.

It is acknowledged that the outputs of sub-Saharan agriculture are diverse. People also make use of other resources in their environment such as woodland and water supplies. Therefore these 'by-products and amenities' must be considered with the conventional concerns of harvestable and marketable yields. There are some SIs orientated towards 'social welfare', but they are vague compared with those for yield and resource base.

The SIs suggested by Izac and Swift are summarised in Table 1.2. The two scales of the matrix relate to spatial scale (cropping system, farm and village) and three categories of 'products' from the system (main products, by-products and amenities). There is no discussion regarding aggregation. That is how the SIs are to be combined to provide some overall picture of sustainability?

Example 2. Tree home gardens Torquebiau (1992) identifies three sets of SIs relating to the sustainability of tree home gardens (Table 1.3). These systems are essentially agro-forestry and at the top end of Figure 1.2. They are therefore commonly perceived as being classic examples of sustainable agriculture. Unlike Izac and Swift (1994), Torquebiau (1992) provides no reasoned argument as to what the time and spatial scales should be, although many of the SIs relate to the garden itself, an entity with a defined spatial boundary. Some consideration is also given to wider impacts outside the system. The SIs are divided into three groups:

1) those relating to the resource base (eg. the soil)
2) those relating to the function of the system (predominantly the use of labour and other inputs along with production)

28 *Visions of Sustainability*

3) those relating to other factors influenced by the system

The SIs in Table 1.3 are technically straightforward, and have much in common with the list provided in Table 1.2, although the grouping is different. It is interesting that the LER for the system is one of the SIs. As discussed earlier, LER is essentially a comparison between intercrop and sole crop production rather than a measure of intercrop productivity *per se*. The rationale for its inclusion is difficult to see as there is no comparison being made with sole cropping. There is a vague and undefined allusion to 'sociological benefits'. The more technical indicators are by contrast strongly defined. Like Izac and Swift (1994), Torquebiau (1992) does not provide guidance on SI aggregation to get an overall picture of sustainability.

Example 3. Sustainability of agricultural systems Stockle *et al.* (1994) derived a scheme based on the quantification of constraints to sustainability in the USA (Table 1.4). These constraints can be thought of as SIs, but the emphasis is on unsustainability (decreasing yield, increasing erosion etc.) rather than sustainability. This is in line with the thinking of Jodha (1989) and others that unsustainability is easier to identify than sustainability. Many of the elements listed in Table 1.4 are reflected in Tables 1.2 and 1.3: productivity, soil and water quality, energy efficiency and economic factors are all present. A new element, 'air quality', is included, perhaps as a reflection of its origins in the USA, while the others given here were designed primarily for developing county contexts. Although there are commonalities between systems, the context of sustainability is clearly important.

Stockle *et al.* (1994) places greater emphasis on social factors relative to the frameworks of Izac and Swift and Torquebiau. 'Quality of life' and 'Social acceptance' appear as major categories, and an attempt has been made to define what they should include. They appear at the foot of the table and apparent the 'ranking' of constraints in Table 1.4 is revealing:

- First: profit
- Second: production (helps generate profit)
- Third: the resource base (underpins production)
- Fourth: wildlife
- Fifth: social factors

Is this perhaps an unconscious reflection of relative importance? Alternatively the ordering may be correlated with the body of quantitative knowledge available in each case, but may also be a reflection of perceived importance.

Despite the usual emphasis on quantification there are no specific suggestions as to how the values can be integrated. Units for all constraints will be different, but what if the values of some constraints are 'good' for sustainability and others are bad? Unlike Izac and Swift, and to a lesser extent Torquebiau, there are no suggestions on spatial or time scales to be considered in the analysis.

Group 2. SI frameworks in practice

Example 4. A Malaysian case study Taylor et al. (1993) focus on five aspects of cabbage production in Malaysia and identify a number of SIs (Table 1.5). This example is different in that sustainability as an approach is used as opposed to sustainability as a property. The emphasis is on practices deemed to be good (or bad) for sustainability. Crop protection (use of pesticide and adoption of IPM) has a particular emphasis in the SI list, and the time scale for change is five years. There is no attempt to include social factors, and instead the list is focused on the technical aspects of cabbage production (at least in terms of crop protection and the soil).

The Taylor et al. (1993) example is interesting as they tried to accommodate an integration of the different SIs. They interviewed eighty-five farmers and gave numerical scores to the SIs for each farmer. The numerical nature of the scores allowed them to be summed, averaged etc. to give a final Farmer Sustainability Index (FSI). The higher the FSI, the greater the sustainability of the cabbage production method for that farmer.

While providing a means of comparison between farmers, this method has limitations. First the scope was limited, focusing on one crop and a few important factors (mostly crop protection and soil fertility) in its production. Secondly, although much effort was made to calibrate the various scores, these were still matters of opinion rather than an objective truth. The weighting of scores has a major influence on the aggregate figure. Valuing some of the SIs differently would produce a different result.

Example 5. A farming systems approach in Burundi A more elaborate approach to measuring sustainability is a technique called 'Weighted Goal Programming' (WGP). An example based around sixteen SIs for human nutrition, three soil fertility SIs and a monetary income SI is provided by Manyong and Degand (1997). The indicators are listed in Table 1.6, and for each a 'required' value is calculated (the reference condition). The WGP calculates a value for 'Z' corresponding to the sustainability of the cropping system. The calculation is based on an aggregation of deviation from what is 'observed' in the system and what is 'required':

$$Z = \sum \alpha \; X \; \text{deviation of observed from required}$$

Where α is a weighting factor that reflects the importance of the SI, and '$\sum$' indicates that the above has to be calculated for all SIs. The less the deviation, the nearer the value of Z is to zero and the more sustainable the system. As seen in Table 1.6 the improved system is deemed more sustainable than the present one, due to its lower Z value.

As with the fourth case study, weighting the importance of the three sets of SIs is inevitably subjective. In this specific case, an equal weighting was employed 'to reflect the same relative importance attached by the farmers to the objectives'. No account is taken of other factors that might influence farmer decision making in agriculture (off farm activities, leisure etc.).

This example further illustrates the central problem of SIs – one can get what one wants by adroit selection of SIs and the way they are measured. The value of Z is essentially a balance between the deviation of the twenty SIs from the researchers notion of what is required. While objective at first sight, there is substantial subjectivity and value judgement resting with the researchers. Measuring the sustainability of each system with a different set of SIs (and different reference points) would provide a myriad of answers and values for Z, and a reasoned argument could be made for each choice.

This example is interesting for a number of reasons. First, it illustrates the drive towards quantitative measurement of sustainability. It incorporates a strong attempt to integrate a wide range of disparate criteria (nutrition, soil fertility and income) covering a range of farming practices. A complex and diverse property such as sustainability is compressed into a single number. Secondly, it emphasises human nutrition as opposed to the more usual stress on crop production, profit, soils etc. Third is the selection of the 'improved' system itself. The researchers made the system less self-reliant by pushing it towards greater, but less diverse, crop production. This resulted in a higher proportion of income being spent on food than the present system, and 'food diversity is obtained through the market rather than from farming'. Yet this loss of self-reliance is not seen to increase dependency, as the staples (such as maize) are still in the system, and production of a high caloric and vitamin content crop (sweet potato) is greatly enhanced. As sixteen of the twenty SIs are nutrition based, the increased production of sweet potato (from 44 m^2 to 1138 m^2) decreased the Z value. This illustrates the difference between sustainability and self-reliance. Finally the time dimension is ignored, and sustainability is simply equated with deviation from required at any one time. The deviation could be reduced or eliminated in a short period by an unsustainable practice. However, changes in Z with time could presumably be calculated so as to develop a vector of system evolution.

32 *Visions of Sustainability*

Summary

Much of the discussion has gravitated towards technical aspects of sustainable agriculture, and particularly the importance of quantification. The relative neglect of the social dimension has been discussed, and illustrated by the SI frameworks presented in this chapter. The examples listed under production (including LERs and the TFP), energy flows (including efficiency), soil factors (cropping intensity), pest control (Biocide index) and the WGP example do not present a biased picture - quantification is at the heart of science in general.

> Most scientists believe that quantifiable facts are essential to rational evaluation of information and accurate descriptions of reality.
>
> MacRae *et al.* (1989)
>
> the major causes affecting sustainability should be reflected in the indicators which should be quantifiable as well.
>
> Jansen *et al.* (1995)
>
> In order for sustainability to be a useful criterion for guiding change in agriculture, its characterization should be literal, system-orientated, quantitative, predictive, stochastic and diagnostic.
>
> Hansen (1996)
>
> constraints that are not quantifiable are of little use
>
> Stockle *et al.* (1994)

Even in the more sociological dimension of sustainability the emphasis may be on more quantitative than qualitative assessments (Wall, 1998). Allied to quantification is the reductionist nature of SIs (for example see Bell and Morse, 1998). It is unrealistic to expect that sustainability, a holistic concept, can be reduced to a few simple and quantifiable indicators. There is clearly a trade-off. Attempting to capture complexity with more and more indicators poses logistical problems and increases the difficulty of quantification and analysis (Flora, 1992). Nevertheless there is something almost counter-intuitive about trying to capture something as holistic and complex as sustainability in a few quantifiable measures despite the fact that indicators have a long history in environmental science, economics and elsewhere. However, what are the alternatives if we are serious about the implementation of agricultural sustainability? Is good practice enough or do SIs offer a better alternative?

To help answer these questions, more practical experience gauging sustainability, including the use of SIs, is needed. There is an abundance of theory and models, but only a few examples of SIs used to measure agricultural sustainability in practice, and even fewer focusing on developing countries. Hence the aim of the research described here is to contribute to this embryonic field by attempting to assess the sustainability (as a property) of agriculture in a rural village in Nigeria. The next chapter will discuss the rationale and methods involved in the research, and much of this was founded upon the ideas and concepts discussed in this chapter.

Table 1.1 United Nations sustainability indicators (SIs) for sustainable agriculture

(a) description

Category	SI	Unit	Desired direction
Driving force	use of agricultural pesticides	t/10 km^2 of active ingredient (a.i.)	decrease
	use of fertilizer	t/10 km^2 of fertilizer nutrients	decrease
	Irrigation	% of arable land	presumably increase
	energy use in agriculture	Joules/tonne of agricultural produce	increase in efficiency
State	arable land/capita	ha/person	presumably increase
	area affected by salinization and waterlogging	ha and %	decrease
Response	Agricultural education	% of Gross Domestic Product (GDP)	increase

Table 1.1 continued

(b) potential problems

SI	Potential problems
use of agricultural pesticides	Ignores differences in relative toxicity, mobility and persistence of a.i.'s. Does not include use outside of agriculture.
use of fertilizer	Does not take account of differences in important environmental and management factors that influence leaching and volatization. Does not include organic fertilizer and crop residue. Does not include applications to grassland. Reliability of data is questionable.
irrigation	Conceptual and methodological difficulties of interpretation. Issues of equity, efficiency and importance are not included.
energy use in agriculture	Agricultural production is affected by other factors than energy inputs. Data for energy use in agriculture at the present time are not reliable.
arable land/capita	Does not give any information on productivity or quality. Permanent crops (tree crops) are not included under 'arable'.
area affected by salinization and waterlogging	Crude measure (problem of definition). Does not differentiate between causes of salinization.
agricultural education	Quality/intensity of education are not included. Assumes a relationship between level of investment and the quality of education provided.

Table 1.2 Some sustainability indicators proposed for Sub-Saharan African agro-ecosystems

Products	Cropping system	Scales Farm	Village
Main-products	ratio of annual yield to potential and/or target yield	profit of farm production	economic efficiency
		ratio of profit to farmer's target income	social welfare
By-products	soil pH, acidity and exchangeable aluminium content	ratio of aggrading to degrading land area	
	soil loss and compaction	nutritional status of household	nutritional status of community
	ratio of soil microbial biomass to total soil organic matter		stream turbidity, nutrient concentration and acidity
	abundance of key pests and weeds		human diseases and vectors
			biodiversity and complexity
Amenities		drinking water quality	drinking water availability
		source and availability of fuel	

Source: Adapted from Izac and Swift (1994).

Table 1.3 Three sets of indicators relating to the sustainability of tree home gardens

Resource base	Function of the system	Other systems influenced by the home garden
rate of soil erosion	inputs/ha from farm and off-farm sources	amount of fuel wood collected from home garden
soil organic matter content	labour requirement /month and /ha	time spent in fuel wood collection
soil bulk density	flexibility in labour requirement	density of wildlife species in home gardens/ha
soil temperature	gender/age in labour allocation	rural industries and their sources of raw materials
under-story level of photosynthetically active radiation	use of traditional knowledge and practices	
albedo	biomass/solar energy ratio	
under-story temperature	cash input/month and /ha	
biomass (/ha)	nutritional analysis of diet provided with home garden products	
under-canopy rainfall	Land Equivalent Ratio (LER)	
under-story air humidity	value of production /month (or year), /ha and /household	
diversity of useful plants and animals	income/ha (/month or /year)	
	sociological benefits	

Source: Adapted from Torquebiau (1992).

Table 1.4 Summary of a scheme based on the quantification of constraints to sustainability

Agro-ecosystem element	Constraint	Trend for unsustainability
Profitability	farm income	-
	dependence on credit	+
Productivity	yields	-
Soil quality	erosion rates	+ (or high rates)
	organic matter content	-
	cation exchange capacity (CEC)	-
	salinization of soils	+
	alkalization of soils	+
	infiltration	-
	water holding capacity	-
	earthworm activity	-
Water quality	nitrate leaching	+ (or high rates)
	pesticide leaching	+ (or high rates)
	nitrates and toxic organics in drinking water	+ (or high rates)
	chemical loading into surface streams	+ (or high rates)
	biological oxygen demand (BOD) in surface streams	+
	coliform counts in surface streams	+
	eutrophication of water bodies	+
Air quality	soil erosion by wind	+ (or high rates)
	fine particulates	+
	odour intensity	+

Table 1.4 continued

Agro-ecosystem element	Constraint	Trend for unsustainability
Energy efficiency	fossil fuel use	+ (or high dependence)
	output/input energy ratio	-
Wildlife habitats	lake and pond sedimentation	+
	wildlife populations	-
	adequacy of wildlife habitat	-
Quality of life	undesirable chemicals in agricultural products	+ (or high rates)
	standard of living in agricultural communities	-
	income level	-
Social acceptance	complaints about food safety	+
	complaints about quality of drinking water	+
	concern regarding long-term adequacy of food supply	+
	complaints about health threats from agriculture	+

+ = increase
- = decrease

Source: Adapted from Stockle *et al.* (1994).

Table 1.5 Sustainability Indicators based on five aspects of cabbage production in Malaysia

Activity	Indicators
Crop protection	implementation of IPM practices
	number of pesticide sprays/season
	changes over last 5 years in application of pesticide
	use of non-chemical means for pest/disease/weed control
Soil fertility maintenance and enhancement	% of total applied nitrogen from organic sources
	number of applications of inorganic fertilizer
	changes over the last 5 years in applications of inorganic fertilizer, livestock dung and other organic fertilizers
	other means to enhance soil fertility and health
Soil erosion control and multi-purpose practices	efforts taken to control soil erosion
	cabbage grown in a crop rotation
	cabbage intercropped

Source: Adapted from Taylor *et al.* (1993).

Table 1.6 Example of the use of Weighted Goal Programming applied to a set of twenty Sustainability Indicators (SIs) developed for a farming system in Burundi

Area	SI	Farming system Present	Farming system Improved
Human nutrition	energy	-	+
	proteins	+	+
	lipids	--	0
	glucides	+	+
	dry matter	-	+
	iron	++	++
	calcium	-	0
	vitamin A	--	++
	vitamin B1	++	++
	vitamin B2	--	-
	vitamin PP	-	-
	vitamin C	++	++
	Ca/P ratio	--	-
	Ca/P ratio	--	--
	lysine	-	+
	methionine + cystine	-	-
Soil fertility	nitrogen	--	--
	phosphorus	--	--
	organic matter	--	0
Monetary income	cash income	--	0
Number of acceptable values (out of 20)		5	13
Sustainability (Z)		386.64	188.9

The symbols indicate deviation from 100% (the required amount):
+ -   = deviation of up to 50% above and below required respectively
++ -- = deviation greater than 50% above and below required respectively
0 = no deviation from required

Z = an aggregate value calculated from the deviations (0 equates to sustainability, and the larger the value then the less sustainable is the farming system).
Source: Adapted from Manyong and Degand (1997).

42 *Visions of Sustainability*

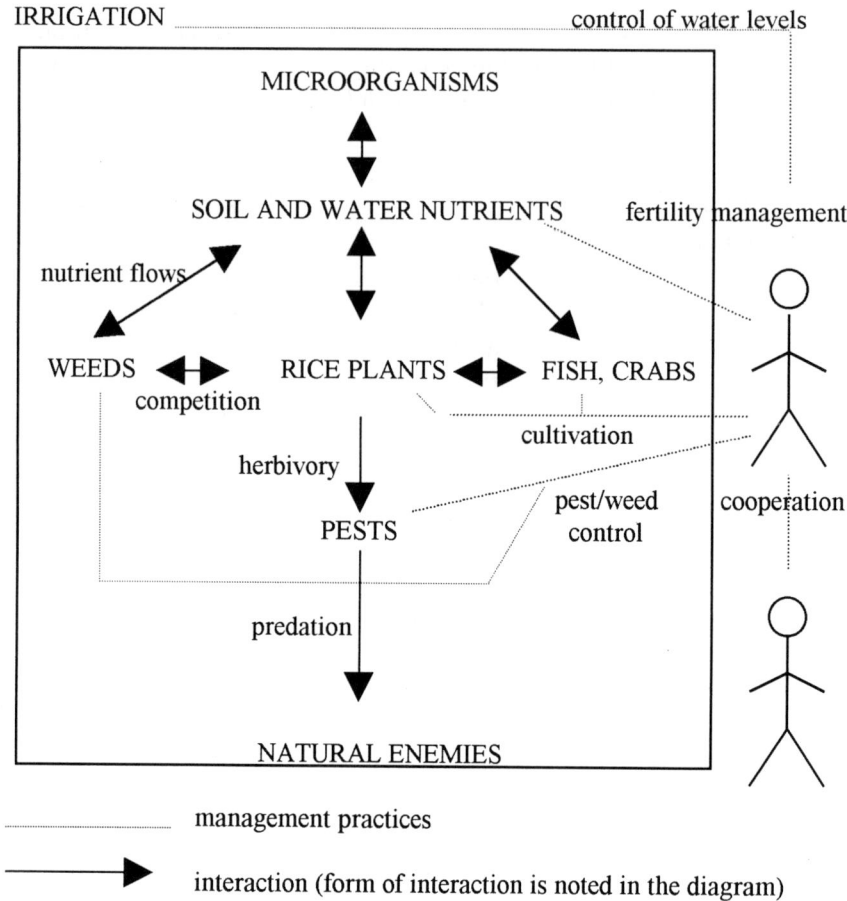

The vision of sustainability as a property depends upon the ability to conceptualise agriculture as a system (in this case a rice field - the box in the centre of the diagram) with definable components and interactions between components. The dashed lines represent management by the farmers, and it is these that tend to be the focus for sustainability seen as an approach.

Figure 1.1   Simplified example of an agro-ecosystem
Source: Adapted from Conway (1986).

		Species separation in	
		Time	Space
Natural ecosystem		no	no

	intercropping systems		
Multiple cropping (> 1 crop/season)	Permaculture	no	no
	Agroforestry	no	no
	Intercropping	no	no
	Relay intercropping	little	no
	Sequential cropping systems		
	(a) >1 crop/season	yes	no
Single cropping (1 crop/season)	(b) 1 crop/season	yes	yes
	Monoculture systems		
	(a) different varieties	yes	yes
	(b) same variety	yes	yes

sole cropping systems

Note – diagram refers to cropping over time on a single piece of land (e.g. a field).

Figure 1.2 Spectrum of cropping systems

44 *Visions of Sustainability*

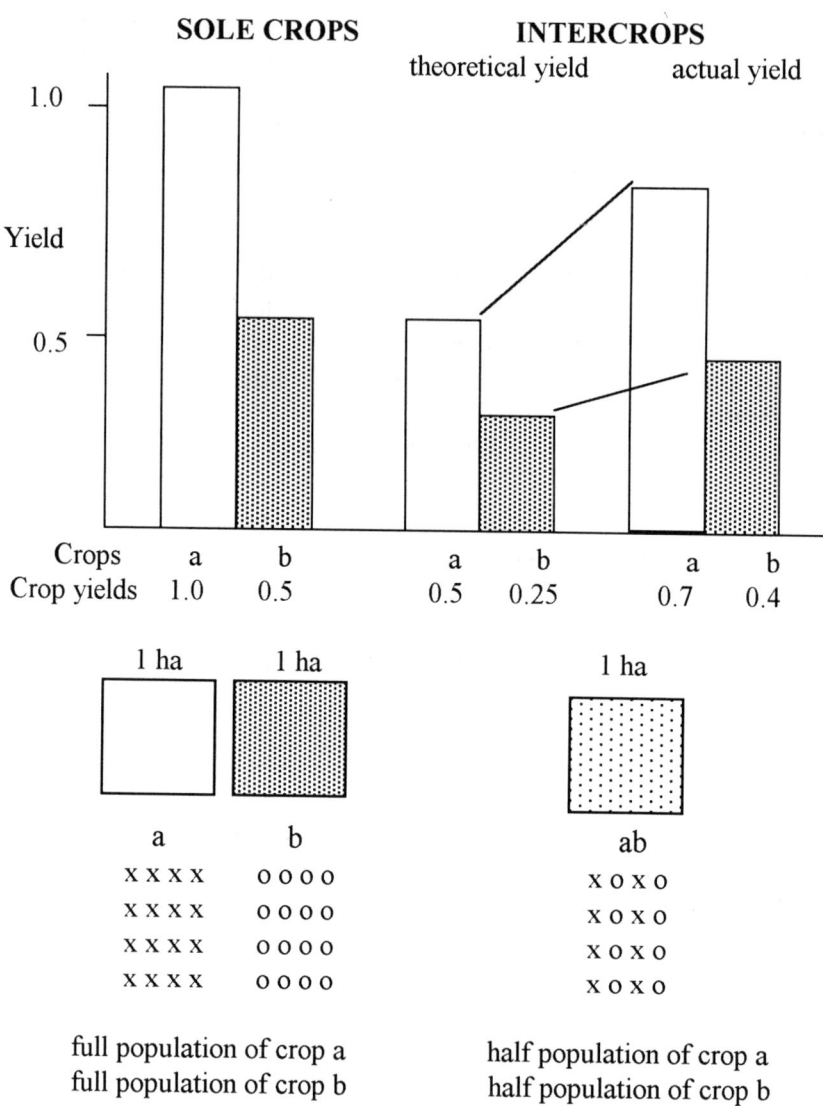

Figure 1.3 Over-yielding in intercrops

Agricultural Sustainability: A brief tour 45

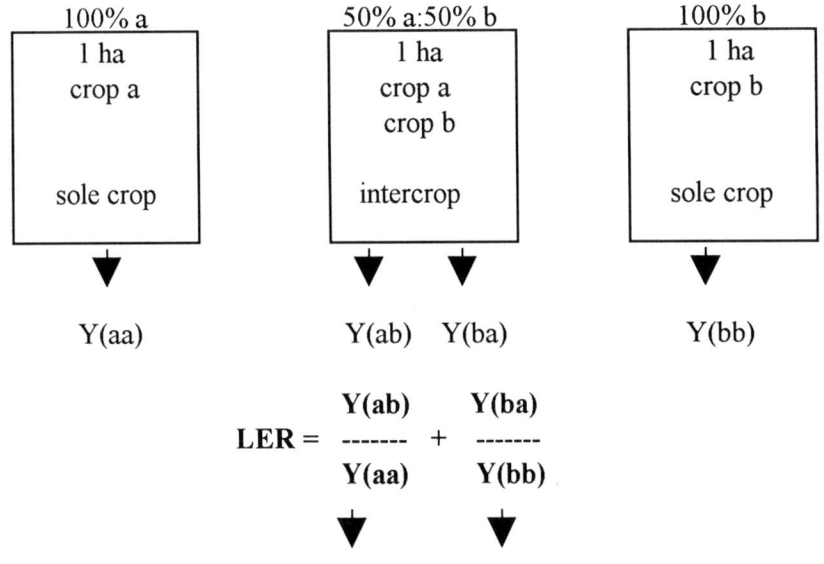

EXAMPLE (sole crop, 'theoretical' and 'actual' intercrop yield data from Figure 1.3)

	Crop	Yield (t/ha)	
Sole crop	a	1.0	
	b	0.5	
		Theoretical intercrop	Actual intercrop
Intercrops	a	0.5	0.7
	b	0.25	0.4
pLER	a	0.5/1.0 = 0.5	0.7/1.0 = 0.7
	b	0.25/0.5 = 0.5	0.4/0.5 = 0.8
LER	a and b	0.5 + 0.5 = 1.0	0.7 + 0.8 = 1.5
		intercrop yield is equal to sole crop yield	intercrop yield is greater than sole crop yield

Figure 1.4 Explanation of the Land Equivalent Ratio (LER) with two crops (a and b)

2 Rationale and Method

Introduction

The previous chapter has set out the current position regarding the measurement of sustainability in agriculture. As will have been noted, much of the emphasis has been on the development and implementation of SI frameworks, and quantification is typically perceived to be of importance. The work described in this book began in 1992, at which time there were only a few published works on SIs relating to sustainability in agriculture though the popularity of indicators was rapidly increasing. During the period of research (1992 to 1998) more appeared in the literature (egs. Torquebiau, 1992; Taylor *et al.*, 1993; Izac and Swift, 1994; Stockle *et al.*, 1994), and whenever possible the methodology was modified to accommodate the new ideas they embodied.

This chapter outlines the methodology employed for the research. The first section will examine the choice of spatial and time scales, and how this lead to a focus on one village and the period between 1987 (structural adjustment in Nigeria) and 1997. These were critical decisions, and need to be fully rationalised. The remainder of the chapter will detail the methods employed in data collection and analysis.

Space and time scales

Though a number of sources were tapped the core elements of Izac and Swift (1994) were adapted to serve as the framework for the last four years of study. These authors rationalised that the 'village catchment' area should be the spatial unit for studying agricultural sustainability, and ten years the minimum time frame (although much depends on the particular SI). They also outlined the SIs to be measured, with an emphasis on production. The Izac and Swift paper was unusual for it included a reasoned set of arguments for all choices. Thorny issues regarding appropriate time and spatial scales for sustainability seen as a property have been largely neglected in the literature except for a universal reiteration of their importance.

Adapting the Izac and Swift methodology for this study was relatively easy as many of their suggestions had already been contemplated in the early 1990s. DDS took the decision to focus on an individual village as the spatial unit for studying sustainability, but the rationale differed from that of Izac and Swift. In a number of respects a Nigerian village provides a logical spatial boundary, and the experience of the authors over twenty years suggested that such a village can represent a microcosm of an area as a whole. However, the main reason for selecting a single village, in this case, was to simplify the logistics. The village was never intended to be treated in isolation as even the smallest village in Nigeria is part of a complex set of economic, social, cultural and political inter-linkages. Village studies (or case studies) have their critics:

> Case studies of change are always open to the charge that they reflect unique historical circumstances and that their complex multiple variables resist generalization.
>
> Netting *et al.* (1989)

Case studies also have advantages. Evidence is collected systematically, the relationship between variables is studied and the study is methodically planned. It is concerned with the interaction of factors and events, and sometimes it is only by taking a practical instance that we can obtain a full picture of this interaction.

The case study method allows the researcher to concentrate on a specific topic and attempt to identify the various interactive processes at work. It would therefore need to examine only a few persons and items. The adaptation of case study to 'community study' was an added attraction as it could provide both an in-depth and detailed analysis as well as the possibility of examining phenomena from which generalisations could be made. There is a risk of over generalisation, but Chambers (1991: 59) feels there is no loss of richness and depth of insight when a researcher resorts only to a questionnaire and/or discussion for data collection. The case study includes all methods of data collection and therefore methods can be selected that are the most appropriate for the task.

In 1992, the original intention was to examine and analyse change since the introduction of Structural Adjustment Programmes (SAP) in Nigeria over the two years 1986/87. It was originally envisaged that the study would take five to six years, and the late 1980s to the late 1990s was a convenient period. Nigerians easily remember the significance of SAP with the onset of retrenchment and rapid inflation. The Izac and Swift paper, with their reasoned argument for a minimum ten year time frame for measuring agricultural sustainability, resonated well with the original intention of the authors, although the reasons for the choice were different. The authors had an additional advantage; they lived in the country and could draw upon their knowledge of the village dating from 1974.

Choice of the village for study

The research was based in the Igalaland area of Kogi State, Nigeria (Figures 2.1 and 2.2), and was designed and implemented under the auspices of a Catholic Mission non-government organisation – the Diocesan Development Services (DDS). DDS began its development activities in Igalaland in the early 1970s, and the wider context of the research will be discussed further in Chapter 3. It should also be noted that the study was intended to provide an evaluation of the work of DDS as well as tackle the research objectives outlined earlier.

A village in Igalaland was selected by applying the following criteria:

1) A long history of engagement with DDS
2) The population had to be greater than 500 but less than 5,000
3) A DDS staff had to be resident in the village or living close to the village

The first criterion was necessary to ensure an adequate level of trust and understanding between the researchers and researched. In the experience of the researchers, Igala people will always want to please by giving answers they feel visitors wish to hear. There is fear of the government institutions, namely the police, local government revenue collectors and sanitary inspectors, and people understandably think that information they provide may be used against them (Richards, 1985).

It was necessary to set a minimum population for the village so as to ensure a good cross section of ideas and answers. Villages of 500 or less are limited in scope, and as people would know each other well there is a possibility of homogeneous answers as they discuss strangers amongst themselves partly to ensure that they give the 'correct' answers. The upper population limit was set to ensure that the village was predominantly agricultural. Villages of 5,000 or greater are more akin to small towns, and agriculture may not be the main activity of its inhabitants. The need for a DDS staff resident in the village was to help provide logistical support, and at the same time act as a key resource person on life in the village.

The application of these criteria resulted in an initial list of thirty-five villages which was in turn used to create a short-list of five villages that also offered relative ease of access. One of these, Eroke, a village that DDS had been working with since 1974 was randomly selected for the case study. 'Eroke' actually comprises three villages, Abo Eroke, Abo Inele and Abo Ojoche which form part of what can be called the 'Abo' group of villages. In order to simplify terminology Abo Eroke, Abo Inele and Abo Ojoche will be collectively referred to as 'Eroke' (Abo Eroke is the largest). The other villages in the 'Abo' group are Abo Atanegoma, Abo Itekpe, Abo Ojuwo and Abo Ogbodo. The 'Abo' group is situated in Imane district, Olamaboro local government area (LGA) of Kogi State (Figure 2.3).

Eroke is a rural village where farmers grow both annual and perennial crops, and its population was thought prior to the study to be approximately 2,000 people (recorded as 1385 people in a 1996 census). The number of households was believed to be approximately 99 (actually recorded as 135 in 1996) divided among 72 compounds. Defining units of study, especially social ones, is problematic (Guyer, 1981), but necessary. The definitions adopted are as follows:

(a) Household. Households are units commonly employed in African community studies.

> The most important economic unit in virtually all West African societies was, and still is, the household.

Hopkins (1973) quoted in Guyer (1981)

A household is a clearly distinguishable social unit under the management of a household head who is nearly always male. With him are his wife or wives and their children, perhaps also with members of the extended family and friends. Due to demographic changes in Eroke, their sons outside the village are not replacing the household heads, and the senior wife often becomes the head of household when the male head dies. Although a household will largely comprise a group of blood relatives, many other relatives live elsewhere. The household shares a community of life in that they are answerable to the same head and share a common source of food (Casley and Lury, 1982: 188).

The problem is that household membership can be very volatile. Individuals come and go at a rate that can be disconcerting for an outsider to behold.

> Although the house and the farms just sit there to be visited and counted, people come and go – on business, on visits, or for seasonal migration. This not only makes data collection practically difficult, but it makes precise calculations of production and consumption patterns in terms of household labour constraints and food requirements problematic.
>
> Guyer (1981)

Given that production, consumption and labour are key considerations in any analysis of agricultural sustainability then the problems are very apparent. Guyer (1997: 25) viewed the use of the household unit as unmanageable at a 'pragmatic level of research method'. Nevertheless, its popularity as a unit has much to commend it, and formed the basis for many of the surveys in Eroke. It is from the household that much of the farm labour is drawn, and for all its mobility occupies a discrete unit of space within the village. But is has to be remembered that the household does not fully encompass the concerns and ambitions of individuals who live there.

(b) Family. A household is not the same thing as a 'family' as the former can comprise unrelated members. In Nigeria, one often talks of an 'extended family' (in contrast to 'nuclear' families) that encompasses a group of people whose relationship may be quite distant. A family can be spread over a number of separate households, and in urban centres members may live by themselves or in more 'nuclear' units.

Family has not generally been employed as a key analytical unit in Africa, largely one suspects because of difficulties with defining the members and dispersion. Yet it is an extremely important unit for its members, and transcends the more diverse membership of households.

(c) Clan. The term 'clan' (olopu in Igala) is used here in the broad sense of a kinship group. Eroke village has seven clans, and the members of each would be related.

(b) Compound. This refers to the buildings and yard where a 'household' lives. A compound can have a number of households resident within it, but each household operates independently of the next, though there is a level of informal interaction at all times. Unlike the other three definitions, a compound is a physical structure not a social unit.

Handles on sustainability: the key areas

Following the selection of Eroke a decision was made to focus on the following key areas in relation to agricultural sustainability:

1) agricultural production and the natural resource base (soils, water, biodiversity)
2) food consumption and use of resources (water, firewood)
3) change in the human resource base (population and cultural change, gender issues)
4) markets, credit and indigenous institutions
5) organisation of labour
6) education and health
7) livelihood and leisure

The first two areas follow naturally from the literature discussed in Chapter 1. The other five areas addressed the human resource base as it was known and understood in Eroke. Prior to the study it was evident that the population of Eroke was undergoing change similar to other villages in Igalaland. This rural area is devoid of industries and opportunities for employment, and lacks basic infrastructure. Migration, particularly of young males, dates back to the turn of the century, and more recently, education had accelerated the drain of both males and females. It was well recognised in Igalaland that migration impacted on agricultural production. The other key areas were selected as they were known to be central to the lives of people within Igala villages.

Broad approach

An advantage of working in Nigeria was the existence of a substantial literature on village/community studies. These tend to be sociological, anthropological and economic rather than focused on the assessment of sustainability as described by the literature summarised in Chapter 1, but inevitably touch on elements considered important to sustainability such as change, diversity, coping strategies and resilience. This literature provided numerous insights into how one could approach a village study for sustainability, and some of the studies pertinent to this research are listed in Table 2.1. The list though not exhaustive, illustrates the wealth of detailed and personalised material available. Some were available at the initial stages, while others were published during the study and provided useful stimuli and inspiration.

Methodologies followed in Nigerian village/community studies were largely of the 'conventional' form found throughout the social sciences. Tools such as surveys, questionnaires, life histories etc. abound. An important consideration for this study was the need to draw upon some of the experience gained in 'participative' and 'rapid' rural appraisals (PRA and RRA respectively). The latter may sound contradictory given the work took six years, but PRA and RRA encompass a whole range of tools and approaches the authors felt would be useful. If the measurement of sustainability was possible, particularly in a development context, then appraisals would inevitably have to be completed in relatively short periods given the universal constraints on resources (primarily financial) unlikely to dissipate in the future.

Of paramount importance was the need to let the village have its say regarding their understanding of agricultural sustainability and elements important to them. The study was essentially extractive, and not intended to bring about change in Eroke *per se*. Naturally, doing the research itself had an influence and this was noticeable during the six years.

There was also the need to engage with the approaches described in Chapter 1. As already seen the emphasis has tended to be on the quantitative gauging of sustainability through SIs rather than qualitative approaches. This necessitated the presence of a strong quantitative element in the study so that comparisons may be made with other SI work. Therefore, many large and formal surveys were included, geared to produce numerical data suitable for analysis.

This research can therefore be said to occupy the centre of three overlapping circles of knowledge and experience:

1) technical discussions of sustainability and SIs (as in Chapter 1)
2) village studies (anthropological, sociological and economic)
3) rural appraisal techniques (rapid and participatory)

It is hoped that this combination is a precursor of greater integration between all three in future literature and plans. However, the main concern of this study was to get a combination of material that would be analysable and yet depict the depth of richness, problems and aspirations of females and males in the community. Experience had shown that there is a difference between what respondents say and what they believe. To overcome this, three approaches were included: naturalistic inquiry, the iterative approach and triangulation.

In naturalistic inquiry, the researcher aims at gaining an insiders perspective by living as the people do and imbibing elements of their existence. However, this takes time to achieve as the essential pre-requisite of mutual trust and confidence, advocated by Casley and Lury (1982: 60-64), takes years to establish. Even after years of contact and good co-operation there are situations where one is still held at arm's length. Naturalistic inquiry is a long-term process which enables the researcher to focus on social and cultural phenomena and gain insights into the 'lived' experience that goes beyond the superficial.

The iterative method incorporates the usual phases of research (setting objectives, data collection, analysis etc.), with the whole process involving continuous feedback loops. Conclusions lead immediately to a validation phase rather than the drawing up of new objectives and the collection of new data etc. Throughout the process, various resource persons and not only the researchers assisted interpretation. As much time is spent on analysis and interpretation as on data collection.

The third approach, triangulation, involved the collection of views from individuals, either in a group or on a one to one basis. Data can be presented to them in order to ascertain a variety of reactions. Reaction to data collected is affirmed or responses may highlight the need for further investigation. Triangulation provides a very useful check as a question can be 'targeted' from a number of different and independent directions. Disagreements between groups and individuals can provide fruitful avenues for further study.

Using the above approaches to the research, data were collected by a number of specific methods:

(a) questionnaires
(b) observation (random visits, transect walking, mapping etc.)
(c) individual and focus group interviews and discussions
(d) key resource persons (including oral historians)

Questionnaires were employed throughout the six years. The researchers were acutely aware of the dangers of bias, and questionnaires implemented by a number of helpers were a means of checking various points arising from discussion or observation. Questionnaires require a careful appraisal preceding their development to avoid embodying only the concepts of the researchers. Chambers (1991: 58), despite his misgivings about questionnaires, argues that they should not be abandoned completely. Previous experience had shown that the questions provided a useful checklist and any information that cannot be gleaned by the interview can be supplemented by other methods.

Observation played a key role in this research. The researchers spent many days in the village over extended periods, and observed many of the processes central to village life (farming, processing, trading etc.). For many visits of long and short duration, the dates and times were pre-arranged. It is conceivable, though unlikely, that the villagers could have arranged what they wanted the researchers to see. In order to counter this possibility, random visits with a seasonal spread were also made to the village. Other tools such as farm mapping and transect walking were also employed especially during the growing season to facilitate a better understanding of the farming system.

As agricultural sustainability could not be isolated from mainstream life in the village, individual and focus group interviews were conducted to discuss special topics or aspects of topics. The premise underlying the focus group method is to generate fresh ideas and insights (Kumar, 1993), which would provide interesting feedback and reaction to any new or unusual information. Such groups are usually homogeneous in composition (e.g. a group of farmers or a group of women traders) and no more than eight in number. The former was necessary because experience, and the comments of many evaluations (e.g. Siebel, 1983: 11) suggested that older men tended to dominate at meetings.

Key resource person interviews were based on individuals or groups, and with people from within or outside the village. These are people in a position to provide the needed information, ideas and insights, and are selected on the basis of their specialised knowledge. Farmers would have more information on specific SIs related to agriculture than would a trader, while women traders have extensive knowledge on important aspects such as sources of credit. A special subgroup of key resources persons was the oral historians. Oral history has not been widely used in the development context as a method of seeking information, even though it is both a methodology and an academic discipline (Cross and Barker, 1953). It was deliberately included in this study as the authors believed it could provide useful insights on change in traditional attitudes and practices in relation to agriculture and oil palms.

There was much overlap between these different techniques of data collection: key resource person discussion sometimes evolved spontaneously into a group discussion as other interested individuals returned from the farm or happened to pass by. Questionnaires designed to target a specific topic often led to discussions on related matters and provided points for discussion in meetings. Throughout, the researchers were very keen to let the research evolve organically rather than impose strict 'phases' within which only one technique was employed.

Key resource persons

These were people who because of their experience and knowledge had more insights than others on the different aspects of village life and Igala as a whole. They included local historians and older teachers who had a keen interest in Igala culture and tradition (Table 2.2). The two issues where the key resource persons proved to be both useful and provocative, were the questions of inheritance and migration. There were many in this village and other parts of Igalaland, who could trace the origins of migration and explain how it had become the practice as from the turn of the century. This provided useful triangulation with formal surveys. Inheritance proved to be a major issue in gender relations within Eroke, and key resource persons outside the village provided many useful avenues for further elucidation.

Crop yield data and farm plot histories

DDS had been involved in crop extension in Eroke since 1976, and since the early 1980s a number of on-farm research trials had been implemented with the major crops (McNamara and Morse, 1996). The analysis of yields from these was an invaluable source of information, particularly when matching yield with food consumption. Details of the methodology can be found in McNamara and Morse (1996). Treatments typically take the form of new crop varieties (NCVs) in some plots and the farmers preferred local variety included as a control. These were planted both as sole and intercrops so as to monitor over-yielding effects described in Chapter 1. In contrast, other studies (Guyer, 1997) have tended to use extrapolations of intercrop yield from sole crops.

The detailed farm structure was determined for six farms in Eroke. This involved the measurement of all plots owned by the farmers, and the history of crops grown in each plot between 1987 and 1998. The latter was established partly by observation (since the mid 1990s when the authors began visiting the plots) and also by recollection of the farmer, his family and neighbours.

Questionnaires

A number of questionnaires were undertaken for this study. Table 2.3a provides a summary of the aims and objectives of the various surveys, while Table 2.3b gives the numbers of respondents. Most of the questionnaires were targeted at a sample of twenty households; a figure assumed to be around 20 per cent of all households and approximately 10 per cent of the total population based on a 1993 census. However, as will be explained below, the 1993 census was inaccurate, and was repeated in 1995. Using the 1995 data, the total number of households in Eroke is 135, and therefore the sample represents 15 per cent of the total. The selected sample of twenty households contained an average of 225 people over the period 1980 to 1998. This is 16 per cent of the Eroke population recorded in the 1995 census. Given the logistical constraints expected with this study, working with more households would have been problematical.

To allow for a possible effect arising from the extensive contact DDS had with Eroke, the households were first divided into two groups: low and high contact with DDS. Criteria for selection revolved largely around membership of the DDS Farmer Council (FC) project. Households which had been involved in the FC project at some time were classified as 'high contact', while households which had never been involved were classified as 'low contact'. Ten households were randomly selected from each. In addition, household selection was designed to ensure that all seven clans were represented in the sample. No differences were expected between them, but it was important to ensure the surveys were inclusive.

A minor change was later made to the sample at the outset of the research. An extra household was added to the 'low contact' category. This was done at the request of the people as they felt that one of the clans did not have enough representation in the sample.

Population census

A census was carried out in 1993, to gain an overview of the composition of the population in Eroke. The census covered all seven clans, and involved counting the number of people resident in each household. The age categories employed were:

Elder: anyone 65 years of age and over
Adult: anyone 18 years of age and over (the age for voting in Nigeria) but less than 65
Child: anyone less than 18 years of age

The adult category of between 18 and 65 years of age is broad, but is a reasonable definition based on consultation and observations. These people constitute the work force. A girl married at 14 or 15 would also be entitled to be called a female adult. However, in order to simplify matters it was decided to put the lower age limit at 18; the age at which people are entitled to vote in Nigeria.

Deciding the upper age limit of an adult male or female was also difficult. The term 'male elder' is sometimes understood to be the same as 'household head', but some household heads may be relatively young. However, no male household head is less than 18 and in such cases the household is female led. In other words, circumstances, as much as age, determine what constitutes a female or male elder. After deliberation with the people it was decided to adopt 65 as the upper age limit.

The 1993 census had a number of problems. Individual family members who lived elsewhere were often included. For example, wives would include husbands, and mothers would include children, who lived outside the village. It was therefore decided to repeat the census in 1995 by which time the researchers were more familiar with the village.

Population change (demographics)

The aim of this survey was to investigate the flow of women and men in and out of households over twenty eight years (1970 to 1998). It rapidly became apparent, especially after the first census, that there was substantial movement of people into and out of Eroke. The survey was implemented in 1995 (just before the census was repeated) and again in 1998, and focused on the sample of twenty households. The technique found most suitable involved writing down the names of the people actually living in the household in certain reference years (called anchor years in the survey - 1970, 1980, 1990, 1995 and 1998). Individuals were classified using the census age groups. Once the population of the household in each of the reference years was established, changes in the intervening period were traced. The questionnaire was completed with the help of two to four respondents from each household, and in some instances, members from other households in the compound were enlisted for assistance. Numerous visits were required to check information.

Village profile

The purpose of this questionnaire, implemented in 1993, was to enter more deeply into the lives of the twenty households. It also allowed a crosscheck of the impressions and insights obtained by means of naturalistic inquiry. A male and female respondent was selected from each of the twenty households (forty respondents).

The particular emphasis was on the productive activities and maintenance of the household, including food preparation, childcare, education, health care and laundry. The household maintenance tasks are essential economic functions ensuring the development and preservation of the human resources of the household and village. A third crucial section was an examination of matters pertaining to access to, and control over, resources. The difference between these two points is critical. Access to resources does not necessarily mean the authority to control them.

Remittance survey

This survey took place in 1996, and was intended to follow-on from the results of the village profile study in 1993 and other discussions. One of the major findings during the first years of the study was the importance of remittances from the 'abroadians'; those living and working outside of Eroke. The aim of the 1996 survey was to determine the types of remittance sent back to relations and friends. A total of forty-eight people from twenty households were interviewed, and these encompassed household heads (mostly men), mothers, sisters, wives, other relations and friends. The categories of remittance were clothes (including shoes), money, food/drink and others.

Migrant survey

This survey also flowed from various group and individual discussions. It looked at the types of skills and agricultural inputs that returning migrants brought back to the village. The survey was relatively small and only involved fourteen male respondents, all of whom had some contact with returning migrants. The respondents were simply asked to list the agricultural

Inheritance survey

This questionnaire was implemented in 1995, and had forty-two respondents. The results of the village profile study suggested that women were now inheriting tree crops, a result that Igala key resource persons outside Eroke refused to believe, as women did not own land. The questionnaire extended to the seven clans, and six households were requested to supply respondents. The respondents were mainly women, but fathers answered on behalf of minors. The result was a coincidental balance of twenty-one male and female respondents. The questionnaire had two sections:

1) inheritance of trees by women from either their father or husband
2) planting of trees (either on the farm or in the compound) by women for themselves

Numbers of trees in each category was recorded for three time periods: before 1970, 1970 to 1980 and 1980 to 1990.

Occupation, income and expenditure ranking

This was administered over two years: 1995 and 1997. The first attempt in 1995 was a pilot study and involved eleven respondents randomly selected. The aim was to get some indication of the main sources of income and particularly the items of expenditure. The second survey in 1997 was much larger, but built upon the results of the 1995 survey. The pilot survey confirmed that it was possible to get data on expenditure, but not income. With this in mind, it was decided to obtain a simple ranking of sources of income and expenditure on a scale of first (most important), second, third etc. Respondents were also asked to rank their main occupations (farmer, trader, blacksmith, herbalist etc.).

Medical treatment

This survey involved twenty one households covering all seven clans. People were asked to enumerate the visits by family members to three sources of medical treatment between 1993 and 1995. The three medical sources were:

1) private clinics/hospitals
2) herbalists
3) DDS clinics/demonstrations.

The focus on medical treatment arose from the first income/expenditure survey which suggested that health care was a major item of expenditure. Respondents were household heads (men) and their wives.

Local inflation survey

This took place in 1996 and asked fifteen respondents to estimate the cost of a number of commonly used local items at three times: 1970 (post Civil War), 1980 (during the time of a World Bank funded development project in Igalaland) and 1996 (time of the survey). Using this data it was possible to estimate the percentage change in prices for the periods 1970 to 1980 and 1980 to 1996. The items included in the survey were chosen by a focus group of four villagers, and grouped as follows:

1) food (maize grain, yam tubers, rice, cassava flour, vegetable oil, palm oil, bread)
2) tools (cutlass, bucket, hoe)
3) building materials (zinc panels, cement)
4) transport (taxi fare, bicycle, motorcycle, speed boat fare)
5) fuel (kerosene, petrol)
6) miscellaneous (shoes, hen, soap, soap powder, baby expenses, Christmas clothes, copy book, pencil)

These figures were then compared with official estimates of inflation provided in International Monetary Fund (IMF) Yearbooks.

Production survey

This survey was repeated twice, once in 1997 and again in 1998. The questionnaire was the same in both cases. The aim was to determine whether components of production (area, yield, quality, pest and disease incidence etc.) for various field crops, trees and animals had changed (increase/improved, same, decrease/worsened) over the ten years between 1987 and 1997/98. In 1997 thirty-two people were interviewed, while in 1998 forty-one people were interviewed. The respondents were taken from twenty-one households, and included males and females from the same household. Also included were questions about water, firewood, pests and diseases and quality of life in general.

Consumption survey

This was implemented in 1998, and took two distinct forms:

1) a ranking of change over the ten years from 1987, in the consumption of certain staples (increase, same, decrease).

2) an analysis of the quantities of the different foods consumed by a household.

Both surveys involved the sample of twenty households, and the respondent was the senior wife in each. The first followed the same methodology described for the production survey, and the aim was to determine whether a trend in consumption existed for the household. The second survey intended to match consumption with estimated levels of crop production in 1997 based on the plot yields. This involved the following steps:

(a) an estimation of annual consumption of foodstuffs, based on known quantities of food consumed/household for different types of meal and the number of meals eaten in a year. The quantities of commonly used foodstuffs used in preparing meals were weighed.

(b) using data from (a) and household size, it was possible to determine (using regression analysis) the quantity of food required for an individual in the year.

(c) plot data were used to determine average yields (and variation) in Eroke. These could be used in conjunction with data from (b) to calculate the area of land required for each crop to sustain a single household member.

(d) farms were mapped so the areas of each crop cultivated by the household in 1997 could be determined.

(e) the theoretical areas calculated under (c) for a household of a certain size could be compared with the actual areas cultivated under (d).

Throughout the exercise, variation around the means and regression slopes were incorporated in the calculations so as to give an 'upper' and 'lower' estimate.

Firewood

The firewood survey was implemented in three phases. The first formed part of the production survey outlined above, and looked at whether availability and household consumption of firewood had increased, remained the same or decreased in the ten years from 1988 to 1998. The second went further, and twenty female respondents (one from each household) were asked for their views with regard to change in:

1) species availability
2) distance from compound to collection
3) whether firewood was purchased in the ten years between 1988 and 1998

The third phase asked a smaller group of ten women (selected from the respondents of the second survey) to rank individual tree species for quality as firewood.

Labour surveys

Two labour surveys were conducted, and both were complex in nature. The first, in 1995 and 1996, measured the labour input for a number of field plots. Labour was divided into activity, gender and type. Activity consisted of land preparation (clearing and ridging or making heaps), planting, weeding, harvesting (including transportation back to the compound) and post-harvest (threshing, winnowing and storage). The type of labour was either household, group, hired help (individuals who work for a specific period of time) or unpaid help. This survey also provided a census of plots under cultivation and allowed a division of systems into sole crop and intercrop. A total of 299 plots were surveyed, belonging to twenty-one men and twenty women (managing 217 and 82 plots respectively). Three enumerators resided in the village for the 1995/96 growing season (cassava takes longer than a year to mature), and recorded labour input for each plot by asking its owner. The plot size (m^2) and crops present were also recorded.

The second survey in 1998 investigated the change in usage of group labour between 1987 and 1998, as well as change in cost. The aim was to determine preferences for the different types of labour, and whether group membership had changed over ten years. Respondents were also asked their opinion regarding the cost of farm work (clearing, ridging and the harvesting of cowpea and bambara nut) in the late 1980s and the current cost in the late 1990s.

Change in education and health expenditure

This survey was intended to look at changes in the cost of education and health as it was evident from earlier surveys, and more especially from everyday conversation and observation, that health and education swallowed up most of the household income. The local inflation survey of 1996 had concentrated on a wide range of items selected on the basis of common need (eg. tools, fuel, food, building materials), and did not focus specifically on education or health care.

A total of ten education institutions (primary, secondary and third level) in the immediate vicinity of Eroke were visited, and key personnel interviewed as to the costs of major items such as registration and exam fees, books, board and photocopying. A similar exercise was completed for three clinics located near Eroke.

Timetable

The study began in 1992 and was intended to run for five to six years. Although a detailed methodology was not established at the outset, the broad intention was to begin with a general analysis of the village and conclude with an analysis of production, consumption and the natural resource base. This order was established to ensure a proper contextualisation of agricultural sustainability within the wider socio-economic and cultural environment. In practice, the general analysis 'phase' took place between 1992 and 1996, while the natural resource analysis happened in 1997 and 1998. There was, however, substantial overlap between these phases.

As soon as Eroke was selected a discussion was held with the chief to explain the purpose of the investigation. Other meetings followed, and the composition of these was similar to other meetings in Igalaland. The men, greater in number than the women were relatively older, while the women looked younger and more enthusiastic. It was at the instigation of the women that the researchers were invited to come to the village and experience life from 'within'.

During these early stages of the study, information was largely collected informally by observation and discussion. Markets, farms and clinics were visited, and the researchers attended prayers and recreation. There was ample time for discussion with the oral historians of whom there was no shortage. Females and males had their stories to tell and were anxious to share them. The different craftsmen (blacksmiths, herbalists, palm wine tappers etc.) came to discuss their role, past and present, and were anxious to show their places of work and explain how their work was done. There was an opportunity to study agriculture in detail, and particularly the role of economic trees (eg. oil palm and locust beans). The differing roles of men and women were observed. It was possible to attend the many organisation and group meetings in Eroke, as observers. Information that needed clarification was later put to focus groups or key resource persons. By and large, the period 1992 to 1996 provided an opportunity to become immersed in, and acquainted with the events that bear witness to the lives of the people in Eroke. Although this was time consuming and expensive, full participation both intellectually and emotionally in the activities of the community is indispensable for any meaningful analysis and understanding of important structures and processes.

The second 'phase' (1997 and 1998) concentrated on changes in the natural resource base between the late 1980s and 1990s. Greater emphasis was placed on questionnaires and the collection of quantitative data. The information obtained in the first phase was invaluable both in the design and implementation of the later surveys.

Analysis and interpretation

For the most part the results of this study have been presented as text, and every effort has been made to preserve the richness found in the community while at the same time illustrating clearly the problems that are integral to life there. Quotations are included though much of their meaning can get lost in the translation. For example, people constantly speak of 'power', meaning strength or ability to perform a particular task, but if translated literally it would change the meaning. To some extent, therefore, we have provided a meaningful rather than a literal translation.

Graphs, diagrams and summary tables with numerical data have been used to illustrate some points. These are included at the end of each chapter so the narrative is not broken. For the most part the analysis involved the calculation of an average along with some measure of variation (standard deviation, standard error or confidence limit). Relatively few statistical tests have been employed, and the majority of these are basic tests such as contingency table analysis of counts (Chi-square test), analysis of variance (ANOVA) and regression analysis. The analysis of counts was used to check for an association between categories. For technical reasons, the convention followed throughout was that if more than 20 per cent of the cells in the table had expected values of less than five, these were excluded from the test. In the tables (mostly in the appendix) the observed values, deviation from expected and the results of the Chi-square test (chi-square value, degrees of freedom and statistical significance) are presented. Regression analysis was employed to determine underlying trends in the data (e.g. the relationship between plot size and cost of labour), and ANOVA to explore variation in data (such as crop yields).

The third type of test employed on some of the rank data was the Mann-Whitney. This is used to test the significance of the difference between two medians. The median is the value in the data set that has an equal number of items on either side, and significance suggests that the group median ranks may be different. The Mann-Whitney statistic is not entirely suited to ranked data, but statistical advice indicated that it would be appropriate.

The researchers naturally endeavoured to interpret the data as objectively as possible. However, being conversant with the village had its drawbacks. It was tempting at times to assume an answer and not ask the question. The opposite was also true, as people sometimes presumed the researchers already knew the answer, and wondered why the question was being asked. In most cases, however, when new information came on stream, the researchers could call on the oral historians and key resources persons, employing their input to check the accuracy and validity of findings. Therefore although the iterative method was time-consuming, it did help to keep interpretations as objective as possible.

The following four chapters will describe the results obtained during the six years of the study. The first two of these will describe and analyse the lives of the people living in Eroke, along with their wishes and aspirations for the future. The remaining two results chapters will look at livelihood in more depth.

Table 2.1 A summary of some of the Nigerian based 'village/community' studies that were used to feed into aspects of the Eroke study between 1992 and 1998

Village(s)	Focus	Author(s) and year
Okiegbo (near Ondo)	Small-scale cash crop Production	Clarke (1981)
Akinlalu and Ilero	Production, income and expenditure of smallholders	Flinn and Zuckerman (1981)
Idere (near Ibadan)	Farm and community Work schedules Change in agriculture	Guyer (1996) Guyer (1992) Guyer (1997)
Ado-Ekiti	Women's work in a textile factory	Dennis (1983)
Abeokuta, Ijebu Ode and Port Harcourt	Migration careers	Peil et al. (1988)
Kofyar homeland (Jos Plateau)	Changed nature of householding	Stone (1998)
	Division of labour in agriculture	Stone et al. (1995)
	Cash cropping	Netting et al. (1989)
	Agro-diversity	Netting and Stone (1996)
	Agricultural intensification	Stone (1997)
Northern Nigeria	Agricultural transformations	Goldman and Smith (1995)
Villages in Anambra amd Delta States	Women in crop production	Ezumah and Di Domenico (1995)
Northern Nigeria	Dry season migration Strategies for coping with hunger	Swindell (1984) Richards (1990)

Table 2.2 List of key resource persons used in the Eroke study (1992 to 1998)

Key resource person list

History teachers

An author who had studied the history of Igala and specialised on traditions

Early migrants who had now settled in Eroke

Retired Ministry of Agriculture staff who had been involved in the introduction of tree crops in Igalaland

Chiefs interested in development

Hunters (retired at the time of interview)

Female and male traders currently engaged in long petty, and short distance trading

Credit facilitators and beneficiaries

Widows

Herbalists

Blacksmiths

DDS extension staff

Table 2.3 Summary of surveys used in the Eroke study (1992 to 1998)

(a) Main objectives

Year(s)	Title	Summary of main objectives
1993 1995	Population census	obtain a broad picture of the population in Eroke
1995 and 1998	Population change	Examine the dynamics of population movement into and out of households
1993	Village profile	list of resources and who has access and control over them
1995	Types of trader and goods traded	List the types of trader and the goods they trade
1995	Inheritance	Examine the changes in the inheritance patterns of women
1995 and 1997	Income, expenditure and occupation ranking	To determine main sources of income and expenditure
1995	Medical treatment	To determine the different forms of health care (already identified as a major source of expenditure): traditional or Western
1996	Remittance	Types of remittance sent to Eroke by the 'abroadians'
1996	Migrants	Skills and agricultural inputs brought back to Eroke by migrants
1996	Local inflation	Trend in local price inflation from 1970 to 1996

Table 2.3a continued

Year(s)	Title	Summary of main objectives
1998 (a)	Consumption (trend)	Trends in food consumption
1998 (b)	Consumption (quantity)	Quantities (especially staples) consumed by households
1997 1998	Production	Trends in production of major field and tree crops. Trends in pests and diseases, and coping strategies especially in relation to soil fertility
1998 (a) 1998 (b)	Firewood	Trends in firewood consumption and availability. Ranking of species in terms of desirability.
1995/96	Labour (plots)	Labour input/plot
1998	Labour (trend in cost and group)	Trends in cost and organisation of labour, especially in relation to the indigenous labour groups.
1998	Education and health costs	Changes in education and health expenditure between 1988 and 1998

Table 2.3 Summary of surveys used in the Eroke study (1992 to 1998)

(b) Number of respondents

Year(s)	Title	Number of respondents
1993 1995	Population census	all households in Eroke
1995 and 1998	Population change	20 households
1993	Village profile	20 households (one man and one woman from each household = 40 people)
1995	Types of trader and goods traded	20 traders (2 male and 18 female) from all the 7 clans.
1995	Inheritance	42 individuals: 21 men and 21 women covering 6 households in the 7 clans. Men answered on behalf of young females in the household.
1995 and 1997	Income, expenditure and occupation ranking	1995: 7 men and 4 women 1997: 25 men and 51 women
1995	Medical treatment	21 households (one male respondent/household; 10 high and 11 low)
1996	Remittance	48 individuals from 20 households. Includes household heads (11), mothers (10), sisters (8), wives (4), other relations (6) and friends (9)
1996	Migrants	14 males that have had contact with returning migrants
1996	Local inflation	8 men and 7 women

Table 2.3b continued

Year(s)	Title	Number of respondents
1998 (a)	Consumption (trend)	20 women (9 high; 11 low)
1998 (b)	Consumption (quantity)	20 women (9 high; 11 low)
1997 1998	Production	19 men and 13 women (17 high; 15 low) 21 men and 20 women (20 high; 21 low)
1998 (a) 1998 (b)	Firewood	20 women (trend survey) 10 women (ranking survey)
1995/96	Labour (plots)	21 men and 20 women (20 high; 21 low) Total of 299 plots assessed (217 male managed and 82 female managed)
1998	Labour (trend in cost and group)	17 men and 15 women (16 high; 16 low)
1998	Education and health costs	10 education institutions 3 clinics

Note

Most of the surveys were conducted with a sample of 20 households (later 21) split into 10 with a 'low' contact with DDS and 10 with 'high' contact (i.e. the household had at least one person who was or had been a member of DDS).

74 *Visions of Sustainability*

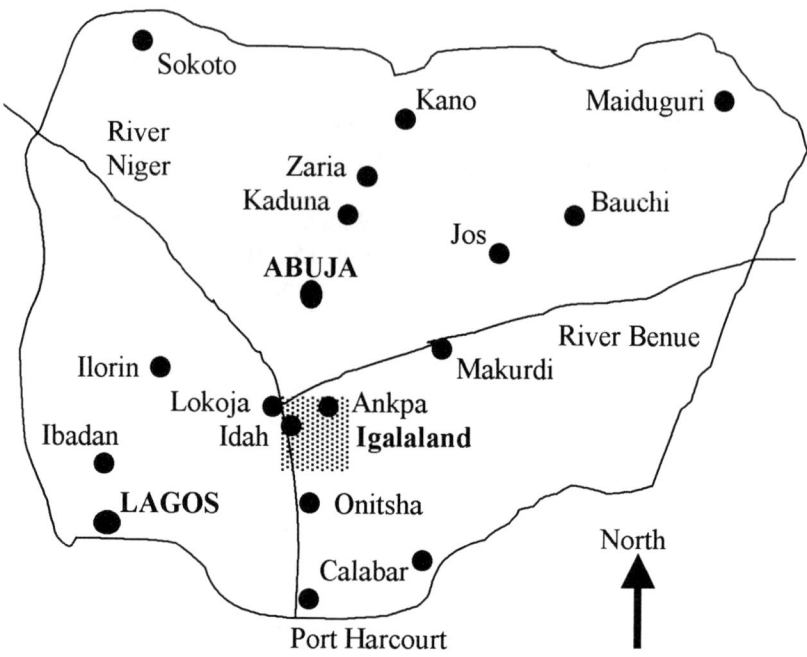

Figure 2.1 Sketch map of Nigeria showing the approximate location of the major cities and towns referred to in the text

Note the central location of Igalaland. Both Idah and Ankpa are relatively small in both size and population compared with the other urban centres shown here.

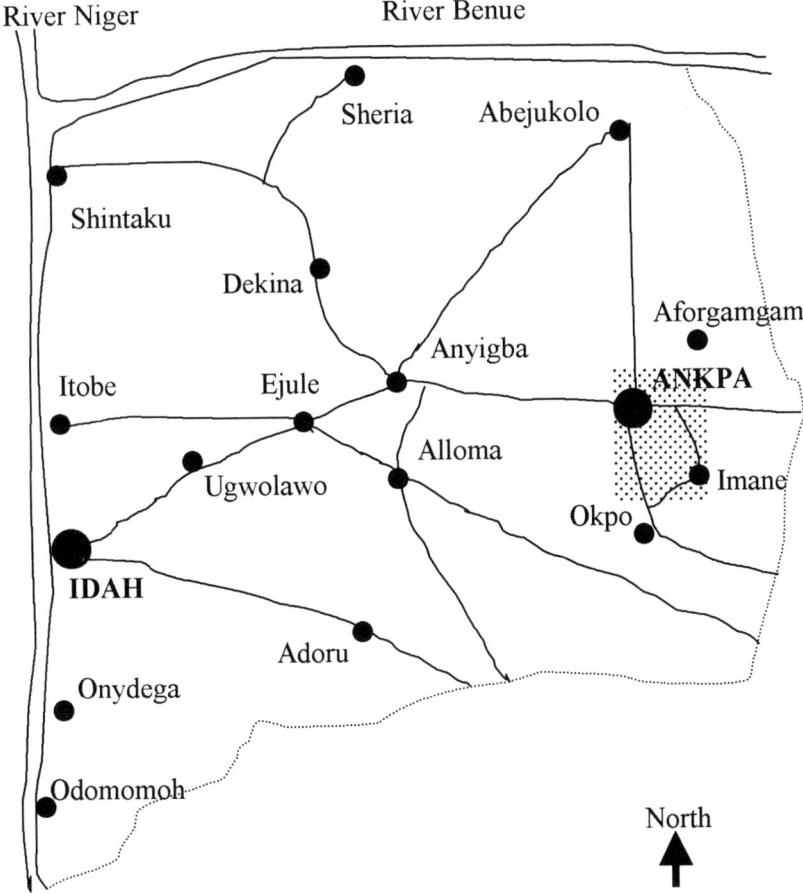

Figure 2.2 Igalaland: major towns, roads (solid lines) and state boundaries (dotted line)

The position of Ankpa in the east of Igalaland gives it a good trading position near the main roads to the south and north of Nigeria.

76 *Visions of Sustainability*

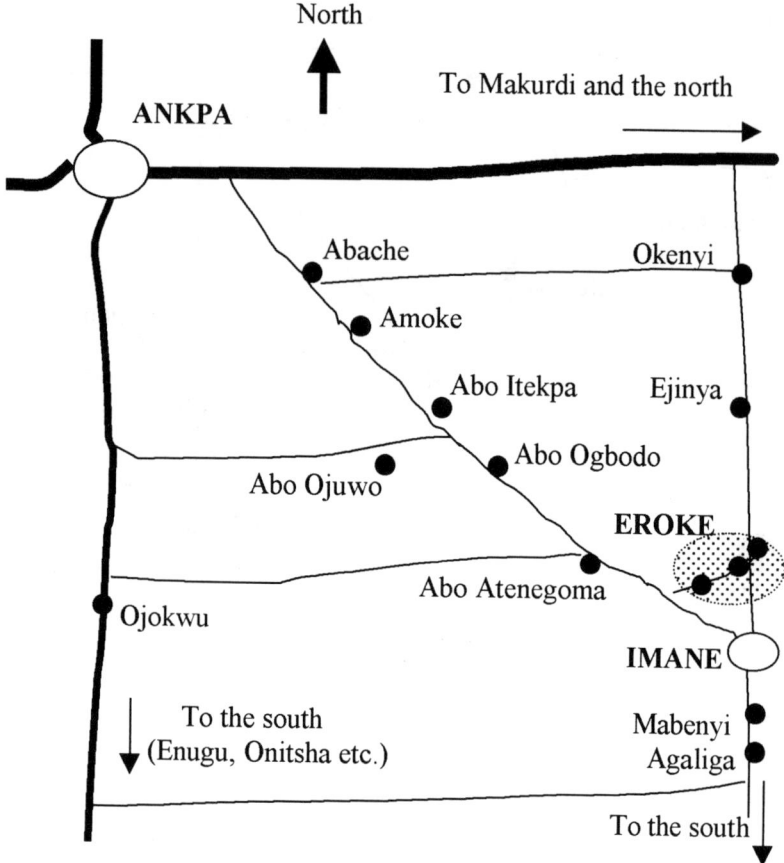

Figure 2.3 Sketch map of Imane and the 'Abo' group of villages

Shown in this diagram are the main urban centres (large circles), villages (black circles), tarred roads (thick lines) and sand tracks (thin lines). The diagram is not to scale, but the distance from Imane to Ankpa is approximately 10 miles (passing through Ojokwu).

Abo Inele, Abo Ojoche and Abo Eroke (shaded area) are collectively referred to as 'Eroke'.
Abo Itekpe means 'in the place of palm trees' (ekpe = palm tree)
Abo Ojuwo means 'on a hill'
Abo Atenegoma means 'on a hill looking around at many things'

3 Eroke Village

Introduction

This chapter looks at Eroke, its people, origins, traditional and political structures and means of livelihood. Special focus is on what they consider important to them, including a tradition of self-reliance, a thirst for education and dependence on the ubiquitous oil palm, without which life would be difficult to sustain. Some of their major problems, such as poor soil, migration and its ensuing loss of male adult labour are examined. These provide the reader with an overview of Eroke, and insights into some of the factors that have influenced their lives. Some themes will be discussed in greater depth in other chapters, but here they are outlined in the wider context.

A period of ten years (late 1980s to late 1990s) has been selected as the focus for an assessment of sustainability in Eroke. In many ways this is artificial. Although in the Nigerian context the choice of a decade following SAP is logical, it is possible to ignore many important changes that influenced Eroke prior to that period. This chapter endeavours to take cognisance of these, and will be covered in greater depth later by exploring some individual life histories (Chapter 4). This chapter and Chapter 4 will attempt to convey the wholeness and diversity of Eroke, while Chapters 5 and 6 will look in detail at production, consumption, income/expenditure and livelihood.

This chapter will begin with an examination of the Nigerian context where the people in this study live and work, followed by a summary of the geography, history and traditional government of the village. The next section looks briefly at population and livelihoods, particularly agriculture, field and tree crops. The oil palm gets special attention given its prominence in the socio-cultural and economic affairs of Eroke. The variety of ways by which livelihoods have 'their being' is also explored briefly. Another important element in any consideration of agricultural sustainability is population change. This dynamic is apparent to all in Eroke, perhaps more than any other change in the last thirty years.

The Nigerian context

Eroke is part of Kogi State, one of the thirty-six states comprising the federal republic of Nigeria. Nigeria became independent from British colonial rule in 1960, and inherited a turbulent and volatile situation due to the size of the country and the vast range of ethnic groups all vying for their share in leadership. A military junta has governed Nigeria for twenty eight of its thirty nine years since independence, and the country has been subjected to great hardships especially since 1985. Its greatest resource as well as its greatest curse, especially since 1976, has been petroleum oil. Prospecting began early in this century, but the oil boom flourished between 1976 and 1983. Unfortunately, the oil riches have made it very attractive for the military to keep power to themselves. The second republic under Alhaji Shehu Shagari began well, but very early into the second term of office (October 1st 1979 to 31st December 1983) the military seized power again on the grounds that the wealth was not being used wisely and the country was in a lamentable condition. This was undoubtedly true, and for twenty months Generals Buhari and Idiagbon attempted to steer the nation back on course, but not fast enough for some of their military rivals who overthrew them in August, 1985. Unfortunately under General Ibrahim Babangida the fate of most Nigerians further deteriorated, and this could especially be seen in the health and education services. Attempts were made to return the country to civilian rule as from 1990, but these were frustrated despite protests from the international community.

Human rights became a major issue in the Babangida regime, and especially under his successor General Abacha. Deplorable situations developed in the oil producing regions, Ogoniland in the south east of Nigeria being an obvious and well-known example. The oil companies developed the industry at the expense of the local population as pollution destroyed their livelihoods. Those like Ken Sara Wiwa and his followers who attempted to defend their position lost their lives. The sudden death of General Sani Abacha in June 1998 and the appointment of General Abdulsalam Abubakar as leader brought new hope. The latter promising a return to civilian rule by 29th May 1999.

this administration has no desire whatsoever to succeed itself and is steadfastly committed to an expeditious hand over to a democratically elected government.

General Abdulsalam Abubakar (July 20, 1998) *ThisDay*

Meanwhile attempts were made to right some of the situations he inherited. Possibly one of the worst afflictions in the country in general, but particularly relevant to areas like Eroke, had been the hike in petrol prices and the irregularities of the black market which were only being controlled in a very limited number of places. This and many of the sanctions imposed on Nigeria in the previous ten years have had profound effects on Eroke, especially on health and education services, supply of fertiliser, poor roads, and a massive increase in transport costs. Increased petrol prices, expensive but fake spare vehicle parts and deteriorating road conditions have posed the greatest threat to traders and road users. Vehicles break down more frequently than before and accidents have increased. Adulterated petrol and diesel have further compounded the situation. The population of vehicles has decreased so this means less availability and reliability for those in need of public transport. The private sector is subject to the same constraints, and apart from the few who will always be better off, the plight of travellers was not good for many years. Were it not for 'mechanics villages' there would be even less potential for travel. Mechanic villages consist of a number of mechanics, each specialised in different car models. They repair and maintain most vehicles and provide old and new spare parts.

General Abubakar, true to his word, handed over to the civilians on the appointed date much to the joy of most Nigerians and the international community. The fourth republic was ushered in as General Abubakar bowed out to General Obasanjo. Twenty years previously Obasanjo had handed over the reins to Alhaji Shehu Shagari. Now General Obasanjo, as the newly democratically elected president, was faced with the difficult task of rebuilding Nigeria. He inherited a battered economy as the military left many economic woes unresolved. Perhaps, the most difficult task is the escalating violence in the south-eastern part of the country which is the major source of Nigeria's oil. Multinational companies like Shell and Chevron, pump away two million barrels a day of high-grade crude from the marshy land around the Niger delta. Tribal leaders complain about the environmental damage that has reduced them to destitution. Militant local groups promise war unless the situation is addressed while oil companies blame the problem on the Nigerian constitution. General Obasanjo visited the region very promptly, but the situation is still seething. Kidnapping has become commonplace and large sums of money are being demanded from the oil companies who are mainly the victims. These are not understood as ransom payments, but seen as compensation for the wrongs endured for decades. Nigeria is still in an abyss, with the oil producing areas presenting one the greatest threats to the country and its new leadership.

Immediately after the election Obasanjo promised his citizenry:

> good governance, poverty alleviation and the eradication of corruption which has been the nation's greatest headache.
>
> Quote from an article (A new beginning) written by Tunde Abaten in *The week*, June 6, 1999

New reforms were promised, and shortly there was a shake up within many ministries and changes within key national institutions including the army, navy, air force and the Central Bank. Women were well represented in the new administration. However, many felt he was only remaking the old order by giving key positions to those active in previous administrations. This pre-occupation is reflected in questions such as:

> Can he stamp out corruption with these men?
>
> Quote from an article (Remaking of an old order) written by Festus Owete in *Saturday Punch*, June 2, 1999

The situation is still far from ideal. The new minimum wage initiated by Abubakar has left private companies and other organisations with no option other than to downsize, and many groups, even state ministries, have been unable to fully implement it. To make matters worse, many traders, on hearing this announcement, increased their prices to take advantage of the prosperity of customers. There is no improvement in the health scene either. Many argue that had health and education services been resuscitated the 'ordinary men and women in the street' would be much happier than with the new minimum wage which has only been effective in increasing food prices.

Import duties have increased by 300 per cent, and there are no signs that the situation will improve. Obasanjo did however stress food security as a top priority. The fuel supply situation improved and available at the controlled price in most places by July 1999. A task force is in place to ensure this continues.

It is against this turbulent and volatile background that agricultural sustainability in Eroke has to be understood and analysed. Although one may feel that the macro-economic picture provides the worst possible backdrop for sustainability, at the 'micro' level one might venture to say that the harder people are hit then the greater the urge to survive and fight back through skilful adaptations.

The Diocesan Development Services (DDS)

The Diocesan Development Services, the development arm of the Catholic Diocese of Idah, was initiated in 1971 as a response to the new role the Catholic Church was expected to play in the wake of the second Vatican Council (1962-65). The church had always been involved in works of mercy such as alms giving, care of the sick and dying, and the education of the poor. But the post-conciliar teachings called for new ways of being more relevant to the world, making it necessary for it to read the signs of the times. While it was good to take care of the poor and the hungry it was now more desirable to investigate the causes of these problems rather than treat symptoms. These ministries would not be neglected, but were no longer sufficient in themselves.

Each diocese throughout the developing world was encouraged to put structures in place that facilitated this new process. Staff from traditional ministries were converted to this work, often without either training or much understanding of what was involved. Social analysis as part of development was a relatively new concept at the time of Vatican II. Idah diocese was slightly more prepared than others in that it had the services of a trained sociologist. Preliminary investigations in Idah highlighted agriculture as a livelihood that needed attention due to its importance in subsistence and its prominence in the Igala economy. Capital accumulation was another major stumbling block to any form of progress, but eventually agricultural finance was brought into being through the adaptation of indigenous saving schemes for this purpose. This socio-cultural and economic adoption, known as Farmer Councils (FCs), has assumed a central place in the lives of many Igalas not only for agricultural credit but also for trade, the occupation most relevant to Igala women. Other DDS projects focus on participatory research, particularly with new crop varieties (NCVs) and more recently with green legume cover crops, nutrition, primary health care and provision of water, usually through rain harvesting. The DDS approach requires that initiative originates from either a community or group and refrains from participating unless the necessary ground work has been done by those wishing to embark on an intervention of their own choice. DDS carefully combines the social dimension with the technical, and participation of intended beneficiaries has been at the heart of all activities. Special concession is given to genuine poverty cases.

Location and history of Eroke

Eroke is located on the main road between Ankpa and Imane in Olamaboro local government area, approximately 28km from Okpo, its headquarters. Annual rainfall is approximately 1400 to 1600 mm, and when approaching from any direction one gets a picture perfect view of the village couching behind a mass of flora that contains no less than sixty-five different species of trees. The village and its natural setting may appear to be untouched, and initially one may be forgiven for believing it to be a slice of 'original' Africa, 'Kiplenesque'. Outward appearances are however deceptive for one discovers that many plants, crops, weeds and grasses are not in fact African in origin but introduced long before Eroke was established.

> Perhaps unfortunately, most outsiders share a vision of African biodiversity as comprising a wealth of large wild animals in surroundings ideally pristine and untainted by 'cultural modifications'.

> Netting and Stone (1996)

This abundance of foliage is also misleading as it gives the impression of fertile soil. In fact the land immediately outside the compound boundaries comprises highly leached sandy latisols, commonly referred to in Nigeria as 'laterite'. Based on fourteen composite samples the properties of the Eroke soil are (mean and standard error in parenthesis): pH 5.64 (0.094), organic carbon (per cent) 1.07 (0.195), nitrogen (per cent) 0.09 (0.016), phosphorus (ppm) 5.29 (1.291), potassium (cmol/kg) 0.1 (0.013), sand (per cent) 77.3 (5.61), silt (per cent) 9.43 (3.466) and clay (per cent) 13.3 (2.91). Soils along the rivers and streams are heavier and richer in terms of the plant nutrients.

The village is on the Igala plateau, with vegetation typical of the region consisting mainly of oil palm forest, some of which has been cleared to make room for agriculture. Due to the sandy nature of the soil, Eroke depends on palm produce to supplement income. Areas left to fallow support scrub trees and bushes as the fallow period is not long enough to allow re-establishment of the natural oil palm forest.

The river near the village is of great value for water collection in the dry season, despite the sharp incline of the banks. The recently constructed water reservoir, which collects rainfall, partly overcomes this problem. The river, which has a very attractive stretch of largely unspoilt natural forest along its banks, provides an ideal location for tree crop nurseries. The Chief is adamant that what remains of the indigenous forest is conserved; bush burning is prohibited and there is an embargo on hunting. Goats are tied during the farming season, and cattle herded by the migrant Fulani people from the north now only enjoy limited grazing rights in the area. These regulations come from an awareness of the need to conserve what remains of the endangered species of plants and animals. The natural forest still has a few large animals, mostly herbivores, and as with most other inhabited parts of Nigeria there are no large predators left in the bush. In the words of the hunters, Eroke is now 'a safe village'.

Eroke is approximately three hundred years old and prides itself on its history and antiquity. There is scarcely anybody over fifteen years of age who is not aware of their roots and kingship origins which can be traced to Idah, the traditional capital of Igalaland. The forefathers of six of the seven clans now residing in Eroke, fled Idah during the Jukun war in the reign of Attah Ayegba sometime in the 17th century. They acquired land and authority there, the latter by overthrowing the local population in Imane, 3km from Eroke. As their numbers grew they spread from Imane to Eroke and have lived there ever since. By then, the population of Imane was mostly Idoma rather than Igala (Idoma border in Benue State is only 8km from Eroke), but over the years they inter-married and each clan can trace its origin to either of the two groups. This is the reason why Idoma and Igala are spoken freely, although the present population regard themselves as Igala, proud of their Idah connection. Citizens of Eroke have the right to become the Chief (Onu) of Imane. The name Eroke confirms that the first settlers likely made a camp at the foot of the oke tree (Ere means foot; Eroke = 'the foot of the oke tree'); a tree that is still common in Igalaland.

Approximately one hundred years ago, the seventh clan (*Ikor*) arrived in Eroke. This group hailed from Ubele, a village 10km north of Eroke in Ankpa local government area. Their plan was to find a place suitable for hunting, though some oral historians also believe it may have been due to warfare in the present Ankpa local government area between 1903-1907. A member of the group eventually married an Eroke woman, a union that made possible the formation of a new clan which was assigned its own land. Such an arrangement is not difficult in Igalaland as men are allowed to inherit land from fathers, grandfathers and the father of wives. This is an important consideration when discussing the sensitive issue of land. Initially this clan was denied the privilege of participating in traditional governance, namely the office of *madaki*. A *madaki* is in charge of a clan, and since a clan without a leader is difficult to manage this embargo has now been removed. However, it is still excluded from the office of chieftaincy of Imane. Although they can retrace their roots to Ubele where they belong to a chieftaincy clan, they are not interested in returning because they are comfortable at Eroke.

Traditional government

In true Igala tradition, Eroke is governed by two chiefs (*gagos*) who are assisted in their administration by seven *madakis* (clan leaders). *Gagos* and *madakis* are always male, with no question of women holding the post. Each clan is known by its ancestral founder's name, and its area of jurisdiction is well defined. Table 3.1 and Figure 3.1 show the clans and their *madakis* in the 'Abo' group of villages, of which Eroke is part. The total *madaki* areas in Abo are fourteen, and Eroke village comprises seven of these. Some large clans can have more than one *madaki*, while smaller clans can share a single *madaki*. For example, in Eroke, the three clans *Ujah*, *Otekwu* and *Igboyi* comprise a single *madaki* area (known as Eroke Ogane) and are united under one madaki. There is no written document indicating the order in which clans are due to hold the office of *gago*. The candidate is first selected by the *madakis* from among the ruling clans. He may not necessarily be a senior or the most educated, but according to oral historians:

> he must have the capacity to hold the people together.

A new *madaki* is chosen from among the males of the clan on the death of the old holder. The offices of *gago*, and *madaki* are for life, and are in fact statutory. As such, both offices receive remuneration from the local government. Local government must approve appointments to these positions, but the Local Government Chieftancy Council (LGCC) must recommend them.

The position of *madaki* is acquired by virtue of one's age and position in the clan. The role they play entitles them to a number of rights, such as palm produce from the clan land together with some locust beans and grains. On ceremonious occasions when animals are slaughtered the *madaki* receives a leg of the animal. They are also entitled to free labour for harvesting palm produce as well as help in clearing, cultivation and weeding for all field crops. A *madaki* collects tax within his own jurisdiction from which he receives an allowance. He also settles petty cases of theft, failure to comply with communal work, divorce and land disputes within his own clan.

The *gagos* are village heads with responsibility for the general administration of the village. They resolve cases beyond the scope of *madakis* for all the *madaki* areas under their care. Each *gago* holds periodic consultation with his *madakis* to ensure the smooth running of the village, and this body can constitute a court to try petty cases unresolved by the *madakis*. The *gagos* attend meetings with the district head. He is also the custodian of all village communal land and cash crops within it. For administrative convenience he avails of the traditional youth leader, known as the '*achiokolobia*', to organise the youth for the communal work required.

There is no financial gain attached to being gago, rather it is accepted as a means of ensuring the continuation of lineage claims to this office and to the office of Chief of Imane. Financial remuneration from local government is N500/month, a sum that is far below the national minimum wage (N3000/month for Kogi State in 1999). Like the madakis, the gago is entitled to palm produce, locust beans, grains and free labour for harvesting palm and cultivation of field crops. If he settles a divorce case where the dowry is refunded, he receives a fee. He can also benefit from dowries paid during a marriage in each of his *madaki* areas.

Kogi State is divided into five Area Traditional Councils (ATCs) which correspond to the five main ethnic groups in the state, including the Igala (Table 3.2). The chairman of each of the five ATCs is a first class chief, and the *Attah* of Igala (the chief of all Igala) is the head of the Igala ATC. The *Attah* of Igala is also the President of the Kogi State Traditional Council of Chiefs, and is responsible for financial allocation to the five ATCs which in turn, pay the salaries of all the graded chiefs in Igalaland, including the Chief of Imane. Within each ATC there are a number of Local Government Traditional Councils of Chiefs (LGTC), and as Igalaland has nine Local Governments then there are nine LGTC's. The *gagos* are paid via their LGTC.

The formal linkage between Traditional Councils and federal/state/local government in Nigeria has interesting repercussions in that change in structures of the latter influence the former. For example, as new states and local governments were created throughout Nigeria a need arose to create extra *gago* areas in Igalaland. 'Abo' previously had two *gagos*, one of whom resided in Eroke, but with the new changes Eroke has been upgraded to two *gagos*, while 'Abo' as a whole has five. The creation of more *gago* areas gives room for new blood, more scope for new initiatives and allows more opinions to be heard.

The population of Eroke

This study took the 1963 census figure as a starting point to measure population as most politically conscious Nigerians claim that the last two censuses (1973 and 1991) were manipulated. In 1993 key resource people in Eroke claimed that its population had doubled since 1963. The official census of 1963 recorded a population of 2154 for the 'Abo' group, and assuming that Eroke had half of this then the village population was 1077 (Table 3.3a). By coincidence, in 1993 a detailed census of Eroke households conducted as part of the sustainability study arrived at a figure of 2154 for Eroke's population. These figures would confirm a doubling of the population over thirty years, but observations and visits to compounds in the mid 1990s suggested that numbers were smaller than was generally held. A decrease, especially in the male adult group often alluded to was noticeable. An additional census was completed in 1995, by that time much more was known of the village and its households. The results are summarised in Table 3.3b. The 1995 census put the population at 1385, and this represents an increase of 29per cent over 1963. The land area of 'Abo' is not known with any certainty, but if one assumes an approximate area of 25 km^2 (villages, farms and fallow) then the population densities were 86 people/km^2 in 1963 and 111 people/km^2 in 1995.

Unfortunately the 1963 census does not provide a detailed break down of the population of 'Abo' making it difficult to determine what changes took place in different age groups. Based on discussions with teachers and others, it is likely that the male adult population was higher in 1963 than 1993, while the number of children was greater in 1993. Adult male mobility and outward migration had increased between these years mainly because of Universal Primary Education (UPE) introduced in 1976. Although UPE was not without its problems, it did successfully create more employment opportunities for those living in rural areas (Csapo, 1983). The increase in numbers of children was likely partly because grandparents increasingly took care of their grandchildren allowing sons and daughters to seek their fortunes outside Eroke. More children also lived due to improved health care.

A detailed study of twenty selected household between 1970 and 1998 showed a net reduction in the population in 1998 relative to 1990, while prior to that the population had been increasing, although the rate had been slowing (Table 3.4). Between 1970 and 1980 the net increase of the household population was 21 per cent, while between 1980 and 1990 it was only 5 per cent. The decline during the 1990s is particularly noticeable in the adult age group (Table 3.4c). Between 1990 and 1998 there was a net loss of thirty eight adults (28 per cent of the adult population in 1990), and was confirmed for the village as a whole by both observation and discussion. This can be expected as this age group would have joined UPE from 1976 onwards. It is noteworthy that one of the intentions of SAP in 1986/87 was to encourage migration back to the villages by promoting rural development, but in the case of Eroke the opposite was true. Based on discussions with key resource persons it was clear that young male adults were the most likely group to leave the village, although in the survey most of the net decline in the adult population between 1990 and 1998 comprised females.

Judging by the figures in Table 3.4 Eroke is a dynamic village in terms of emigration and immigration. The majority involved in this movement were female, and in the case of emigration it tends to be those who married and moved to another village in Igalaland. Male emigrants, on the other hand, tend to go 'abroad' (i.e. to the major cities in Nigeria) and are known as 'abroadians'. Far from cutting ties with the village, they contribute to its sustainability in numerous ways, a point that will be discussed later. Another notable feature of the figures in Table 3.4 is the relative stasis of the birth rate between 1970 and 1998. In the three decades (1970 to 1980; 1980 to 1990; 1990 to 1998), the birth rate was between fifty two and fifty four children for all twenty households – a remarkable consistency. The main determinant of birth rate, of course, is the number of female adults in the households, and given that more female adults than male left the village between 1990 and 1998 this consistency may perhaps be even more surprising. The explanation was that many of the women coming into Eroke were already pregnant and returning to give birth. Striking in Eroke nowadays is the large number of children attending nursery and primary school. This says much about the dedication of parents to their children's and grandchildren's education.

Livelihoods

The traditional occupation in Eroke is farming. The following quote from a leading woman in the community sums up this attachment:

> They preferred farming to any other thing. Their very own sons were not sent to school because farming was too important.

This refers to the central role farming played in their lives forty years ago. To-day men still do most of the heavy work (land preparation and weeding), while women look after the post-harvest tasks considered by them to be lighter work. Besides farming each family has a trade which is passed from one generation to the next. These include hunting, black-smithing, wine tapping, palm bunch cutting, cloth dying (now extinct), oracle consulting and rain making. There are also those engaged in traditional health care, mainly as herbalists and birth attendants, while others are teachers, drivers, carpenters, masons and lady tailors. Fabricating musical instruments is also an important occupation both for leisure and the economy as a whole.

Another important source of supplementary income for men and women is obtained from the sale of artefacts used in local medicine. Of import are tortoise, snake skins and heads, feathers and heads of rare birds, bones of rare animals as well as barks, roots and leaves. Those with the privileged information required for these formulations guard it securely until it is time for them to pass it on to a son or daughter. Only certain families enjoy such rights, but they apply equally to men and women. Income is also derived from the weaving and sale of baskets and mats useful for doors and beds, the manufacture of traditional fans, domestic utensils such as the mortar and pestle, wooden stools, plates, spoons and musical instruments such as drums.

Goats, hardy ruminants requiring little management, are kept and sold only in emergencies when ready cash is unavailable. They are occasionally killed for festivals thereby saving money otherwise used to purchase the animal. Dogs generate income as they are in demand as watch dogs and for hunting. A litter of puppies can enhance household income as the animals are sold before their food needs are too great.

Women have their own farms, independent of the household, where they typically plant cassava, benniseed and bambara nut. These are mostly for sale though they can retain benniseed and bambara nut as ingredients for soup. All Eroke women are experts at processing palm produce, especially oil, which during off-farm seasons becomes a major occupation. The cracking of palm kernels is a tedious process, normally carried out on a part time basis with help from female and male elders and children. Other than trading in herbs, food and grains, the processing of palm oil and its allied products form the major part of female income in Eroke. A few women engage in preparing bean cakes (*akara*), steamed beans (*moinmoin*) and unsweetened corn flour or custard (*akamu*). These are sold as snack foods or as supplements to breakfast, a lucrative business. Porridge made of millet, guinea corn or maize (known as *obiolo*) is also sold as breakfast to farmers working in the field, often by their wives. Although hunting is prohibited, some game is trapped by men and given to their wives to prepare as pepper soup. This is a delicacy for the family and fetches a high price everywhere.

However, it is as traders that the women of Eroke are best known. This is partly due to its proximity to the major town of Ankpa, a strategic geographical location for trade (see Figures 2.2 and 2.3 in Chapter 2).

Although small by Nigerian standards, Ankpa is a nodal centre and unlike many other Igala urban centres it is well serviced by good roads. In addition, Ankpa-based entrepreneurs own more than a third of all vehicles in Igalaland. Ankpa would have an influence on Eroke similar to what Guyer describes Ibadan having on Idere (Guyer, 1997), only on a far different scale. This is a major advantage to traders there. Food and agricultural produce are transported from Ankpa to all parts of Nigeria, and other commodities fill the lorries on the return trip. As a result Ankpa town is a hub of activity, with much of the hustle and bustle around the market place and motor park.

Eroke women traders have three classifications for themselves:

1) local or petty traders
2) short to medium distance traders (Onitsha and Oba)
3) long distances traders (north to Kano, Jos or Maiduguri)

The long distance traders normally sell palm oil, processed cassava, oranges, mangoes and locust bean, and return with onions, tomatoes, dried fish, beans, cowpea, millet and white guinea corn. These they sell either wholesale or retail in Eroke or other nearby markets. Short distance traders are similar to the long distance traders in the kind of produce they buy and sell, but tend to trade in markets closer to Ankpa. Petty traders sell ingredients for soups, processed foods and washing powder, mainly within Eroke.

Black-smithing is essential to Eroke's economy, but important as the trade may be, there is only enough work for one family. This family hails from Ajobe, a village north of Eroke in Ankpa local government area. Products include kitchen ware and hammers, as well as the implements used for farming, hunting, palm bunch cutting, palm wine tapping and even specialist equipment for herbalists in Eroke and surrounding villages.

There are many herbalists in Eroke. They cure diseases, set broken bones and provide potions that help women during pregnancy and childbirth. They also treat cases of infertility and prescribe for families that lose children prematurely, a common phenomenon known as *abikwu*. To overcome the problem, which can be attributed to either parent, a course of special treatment is mandatory before a new pregnancy can be contemplated. The remuneration of herbalists is small, consisting of an initial down-payment and another token of appreciation if the treatment is successful. Over-charging is considered an offence to the source of the gift, which could be revoked if it angered the ancestors. Some inherit their profession, but others like Asana (Chapter 4) received hers in a series of dreams.

Agriculture

As in many other Igala villages, agriculture is much revered in Eroke and farmers grow a variety of both annual and perennial crops (Tables 3.5 and 3.6). The first concern of every household is to provide enough food for the year, which must be planned for and achieved during the growing season. All agricultural operations, including those of post harvest, are done manually. There are two growing seasons referred to as the early and late seasons. The early growing season coincides with the first rains (usually in April), while crops that require later planting are reserved for the late season. The rains normally end in October, but there may be a short break in early August. Farm activities, commencing with land preparation, take precedence over all others activities. The rainfall can be erratic at that time of year, making crop establishment risky. It is also during this season that many fruits are harvested. The palm fruits are harvested a number of times and require immediate processing. Other tree crops harvested at this time are the locust beans, but their processing can be postponed until the palm processing and sowing of the early crops are over.

Field crops

The annual crops include the major crops of West Africa, and consist of various species of legumes, root crops, cereals and vegetables. The legumes mostly comprise groundnut, cowpea, bambara nut, pigeon pea, and more recently a limited amount of soybeans. The principal root crop now grown is cassava and many new varieties (NCVs) can be seen in Eroke. Yam has always been a popular food, and is still grown although yields are now poor except in the moisture retentive compounds where it is often possible to produce a tuber weighing 22kg. Igalaland is thought to be one of the areas of domestication for white yam. The chief cereals are guinea corn, millet and maize. Rice was introduced in the last ten years, but not common on the free-draining sandy soils. Most of the rice varieties are upland, and grown mainly in the small riverine area close to the village.

The vegetable range has increased in the past years, due mainly to contact with Igbos (an ethnic group to the south of Igalaland) and improved knowledge of nutrition. Spinach is produced both for consumption and sale. Egusi melon has always being a favourite crop, suitable for soups and cash crops, but unfortunately this too needs fertile conditions to yield well. Peppers are produced in great quantities. Okro (lady finger), is still plentiful and grown near the compounds. It is sold either fresh or dried, the latter (*ijagada*) is used abundantly for soups throughout the dry season. Tomatoes grown in small quantities, are a luxury used almost as rarely as meat. They are eaten as part of a stew that accompanies rice dishes. Onions are not grown, and like tomatoes are used as a treat on special occasions.

A household in Eroke can grow between fourteen and eighteen crops each year, typically as intercrops but also as sole crops. Many of the crops are planted in the early season, while the late season crops, mainly maize and cowpea, are sown at the end of July or August often in land that had an early season crop. Short duration crops planted in the early season, such as legumes, maize, and the early millets, will normally be harvested from July onward. Long term crops, such as yam and cocoyam, are harvested at the end of the rains, while cassava may take eighteen months to two years to mature. The first crops to be harvested are a great relief to families as the stored food supply of the last season has often been exhausted by that time.

Tree crops

The perennial tree crops have traditionally been oil palms, the two species of locust bean, the bush mangos and kola nut. More recently, fruit trees such as mango, citrus, pawpaw, cashew, guava, banana and plantain have assumed a major economic role. Many of the trees are NCVs introduced by DDS and the Ministry of Agriculture; NCVs currently constitute 20 per cent of all oil palms. Planted in the last ten years (late 1980s) they are now regarded primarily as cash crops. The NCVs give increased yield of high quality fruits, and this results in a substantial surplus for sale. In NCVs of palm, bunches develop at the base of the tree eliminating risks for cutters and costs for owners. Major changes have also taken place in recent years because of migration and education. Those previously unaware of the value of fruit trees as a cash crop began to plant and sell the fruits to women who marketed them. This new awareness resulted in an increased commitment to better care and maintenance of fruit trees, which in the past were often neglected. Many requested help and training from DDS. In the past these trees were planted, owned and harvested by men only, as rights of ownership over economic trees derive from ownership of the land. However, based on their new awareness of these trees for commoditisation and monetisation, men began to plant for their daughters and wives.

Quality fire wood is now scarce in Eroke as women have to travel long distances to source it or else use fruit trees or even older oil palms for their normal requirements. One would have expected them to compensate for this with fast growing trees such as teak and gmelina, but it may be they are using up stocks of older palms before embarking on any more new ventures.

Land and labour

Land is clearly an important resource in Eroke. There are two categories:

1) communal land
2) clan land

Most clan land is close to compounds, with communal land further away. This latter point is important in an area where the male adults are advanced in years. All male indigenes are entitled to both types, while strangers have rights only to the communal lands or a portion of land from a host. Strangers are not allowed to plant trees on the communal land, but indigenes so doing have rights to that land as long as they live in the village. Those farming on the communal lands pay tribute in kind to the village elders and the *gagos* and *madakis*.

Division of labour between men and women is a common feature of Nigerian agriculture (see for example Stone *et al.*, 1990, 1995), although this is not a standard picture (Ezumah and Di Domenico, 1995). Eroke follows the general pattern in Igalaland. Women are not generally involved in land preparation, but given the present composition of the adult population, it is now common to find women, especially widows, who occasionally till and ridge their own plots. Such women are not averse to growing a few yams for their family, and neither is there protest when it happens. Men are free to engage in all farm activities and generally free to grow the crops they wish.

Vegetables however, were not regarded as a man's crop, but it is rumoured that this changed in the last ten years and they now sow mainly okra, tomato and spinach. This is the outcome of commoditisation and monitisation of vegetables. It is possible that vegetable growing was perceived as beneath male dignity, and it was not unusual for those who consumed vegetables on a large scale to be given nicknames. Women on the other hand, were always more anxious to produce nutritious food, and quickly took the opportunity to grow vegetables regardless of such consequences.

Men do most of the weeding, but household members are involved in harvesting, including the elderly and children. The harvested produce is conveyed to compounds where it is either stored or processed. Men do the storing, but only after the harvested crops have been dried thoroughly under the careful management of the women. Men, women and children peel cassava tubers, and once peeled the cassava can either be dried and stored immediately or soaked in water for three to four days to remove the cyanide before drying and storage. The shelf life of dried cassava is five to six months, and before eating, it has to be pounded or milled. Yam is consumed soon after harvest, although the seed tubers are stored on the farm in field clamps. Grains are dried and processed on the farm or near the compound prior to storage. Both types of locust bean (*ukpehi* and *ugba*) are harvested by men and brought to the compound by women and children who also dry and process it. One of the locust beans (*ukpehi*) is taboo in the village, and not consumed although is sold as a cash crop.

Indigenous group labour

Group labour is important in farm work in Eroke as throughout West Africa, and invaluable as a means of coping with timing constraints and the tedium of working on an exacting task by oneself (Richards, 1990; Stone *et al.*, 1990; Guyer, 1992, 1997). There are three forms: the *ayilo*, *adakpo* and *ogwu*. These are similar, and depending upon the wishes of the group the three can interchange even within a single year or growing season. Membership of either group is entirely voluntary, and a group can comprise a minimum of three and a maximum of fifteen members.. These three groups can be either entirely male or female, but never mixed. The main features of the three types of group are:

1) a*yilo* – rotational system between members without payment
2) a*dakpo* – rotational system between members who can work for payment outside the group.
3) *ogwu* – payment given to complete a task in a given area of land

The *ayilo* is a rotational system, and consists of a group of farmers who work for a definite period of time on the farm of a group member. When finished they move on to the farm of the next member after a short break of a few days until tasks are completed on all farms. There is no cash payment, but the member whose farm the group is working on is responsible for providing food, usually in the form of *obiolo* (guinea corn, millet and maize porridge). Female *ayilos* do planting, weeding and harvesting, while male *ayilos* can be involved in all farm activities. This is a common division of roles in West African labour groups (Richards, 1990). Membership of an *ayilo* is fairly permanent, some drop out because of health or age, but never before completing their obligations. If a member is absent because of sickness, then she/he has to find a replacement or make arrangements to work on another day.

The *adakpo* is different to the *ayilo* in that the group hire out their labour for a fixed sum (cash or credit) to either a member of the group or an outsider. The recipient, that is the person hiring the group, may work with it to ensure good value for his or her money, and has to provide a breakfast of *obiolo*. An *adakpo* can be formed in its own right and without *ayilo* obligations.

The o*gwu* resembles the a*dakpo* as it involves cash payment, but is not rotational and does not necessarily involve a 'group'. An *ogwu* may consist of only two people, and it is seen as cheaper than the *adakpo* where many have to be paid. It also offers more flexibility than the others. Payment is made to both the *adakpo* and *ogwu* to complete a certain task rather than on a daily basis. The agreed price will reflect the time needed to do the work as well as the type of work – some tasks are more strenuous than others, but the incentive is for the group to complete the task. Farmers can also hire labour for a defined period of time, but this is not so popular as it is less efficient than group labour (Richards, 1990).

Both men and women can hire *ogwu* and *adakpo*. Although the rotational labour groups are a useful strategy, the chronically sick in Eroke do have problems with participation in much the same way as Richards (1990) describes for other parts of Nigeria and Sierra Leone. This point is particularly important given the ageing nature of Eroke, and will be returned to in Chapter 6.

Another important form of group labour is the age grade system. These groups are involved in maintenance of the village rather than farming, and again the age grades can be either male or female in membership but never mixed. Age and gender determine the group task, and younger groups and women are not called on for the heavier work. There is no remuneration for such activities, and it is seen primarily as a duty to the community. An exception to the latter may be road maintenance when visitors or locals may be given a token of cash in appreciation.

The oil palm

Perhaps the most graceful tree in the Savannah belt of Nigeria is the oil palm, and it is little wonder that Igalas have chosen it as their emblem. Indigenous to the locality, it continues to be a major source of human and economic subsistence. More importantly, it is a renewable natural resource with many uses, and Eroke is blessed with numerous old and new varieties. Of all the useful products, palm oil ranks first in importance followed by palm wine. In addition, the leaves are used for roofing, fencing and brooms, and the mid-rib is used for baskets, rope, mats, and cages for birds and domestic fowl. The oil palm stem is highly treasured for building and fuel wood, especially as the availability of more favoured trees is declining.

Palm oil, deep red in colour, is extracted from the palm fruits and widely used in food preparation and medicinal formulations. Palm oil is also used in soap-making. Chaff from the oil production is useful as fuel, tapers and traditional candles. Even the waste left after the decantation of the oil (known as *ite)* is a delicacy eaten with boiled or roasted yam. The kernel too has many uses, especially since it can be converted to another oil known as palm kernel oil (popularly called PKO). Women use it for their hair, and herbalists recognise it as a component useful for the treatment of many sicknesses, especially ear ache, back pain, stomach problems, snake and scorpion wounds and even for treating 'dislocations' (bone-setting). The hard cover of the kernel (the cone or pericap), is a good fuel but also often used to control erosion around compounds. Nothing goes to waste, and even the residue left after oil is extracted from the fresh bunches can be combined with the decayed male flowers to produce organic manure rich in potassium.

Wine tappers extract palm wine from the male flowers. Men are responsible for the cutting of bunches and tapping the palm wine, risky with local varieties as they must climb the trees. Men, women and children are all permitted to partake, although the 'sweet wine' (newly tapped) is for the women and children as fresh wine has no alcohol. It is said to improve lactation for mothers. Men prefer the maturer brew, and the alcohol derived from palm wine when distilled and purified produces a dry gin (known as *ogogoro*) which is highly intoxicating. If well prepared *ogogoro* can have an alcohol content of 98 per cent. Apart from its wide acceptance in social gatherings, palm wine is high in nutrients such as yeast and vitamins.

Ownership of palm trees takes two forms: ownership by inheritance and ownership by trust. In the first case, the *madaki* is automatically the caretaker of the trees until his death. During his lifetime he can decide what he wants to do with them. It is also possible that on the death of the *madaki* the trees are shared between all male members of the clan. Attempting to divide such property always causes rivalry, and is a common cause of conflict. Dividing and subdividing of trees on a continuous basis among clan members can mean that inheritance may be so small as to be economically insignificant.

The second category relates to palm trees held in trust. Every taxable male adult in Eroke is entitled to palm trees on the land they own. Once in his possession he has the right to use them until his death or when he emigrates. Occasionally adjustments have to be made, for example, when an individual returns to live in the village. In this case, should none be available, those currently enjoying an allocation can be requested to release some trees for his use. On the death of a person enjoying such assets, the trees are re-allocated either to those in need for the first time or to those who already have them. The number of trees an individual owns is subject to change, increasing or decreasing depending on these circumstances.

Marriage, gender roles and social class

Approximately 80 per cent of Eroke men marry from outside the village, a custom quite different from other Igala villages. A possible explanation is that Eroke men have always married because of their love for a girl, rather than by arrangement. This practice may be surprising as there are more women than men, but as the overall population is small this may have meant marrying a cousin, a practice taboo in the village. In typical Igala tradition, the man is head of the household in Eroke and dictates the order of events. In the past such a stance automatically made women second class citizens, but over the years cultural adaptations have brought change. Because of their involvement in trade, women are now acknowledged as making a major contribution to the overall income and welfare of households. They have a strong voice and are seen by men as 'partners in progress'. Their economic power and freedom can be seen in the number of houses built and owned by females. Such buildings are readily identified as a piece of metal ware (basin, plate or tray) is attached to a prominent part of the roof.

Polygamy is still the norm in Eroke, and there are reasons for the custom. Extra hands are needed for domestic and farm work, and the desire for children, especially males, is still strong. Because of polygamy, female relationships in households are often competitive and rivalry is rampant. This can be positive for the household as it sometimes stimulates economic progress, but it also creates tension. To deal with this, husbands provide alternative accommodation for their wives, usually in their parents' compounds where they have access to resources. Where the husband can keep the entire family as a unit, the senior wife is regarded as 'mother' by all the children and the other wives respect her as such. If separation of wives is necessary, the father lives with the senior wife who must accept the children of other wives as her own. When children from the other wives are weaned they are brought to her. This is a means of ensuring solidarity among the children. It is easier for the wives who live separately to retain their daughters as the father has more interest in his sons. Such an arrangement guarantees the three main concerns fathers have for sons:

1) they know the extent of their father's assets
2) they imbibe their fathers values
3) they learn the family trade

Girls do not need to take on board the family values as they marry and leave the household. Nevertheless, the mother recognises the need for girls to be 'economically viable', and initiates them into all skills and trades of value, regardless of their level of education. Even a tertiary graduate learns palm processing. At present, all children of school going age are in school in Eroke. Their progress through formal education is determined by individual motivation and application, but is also influenced by their parents ability to finance them. Fathers pay fees though mothers may help. However, while mothers rear and train their daughters, the father nevertheless receives the dowry upon their marriage. Most mothers are consoled by the fact their daughter will always be helpful to them.

Resource maintenance is gender determined. Some resources are clearly the exclusive reserve of men while women maintain responsibility for others. Property and assets of high value such as roofs of houses, construction of water channels and mud walls are the responsibility of men. Men carry out these jobs themselves, but also hire skilled labour for specialist tasks such as the repair of kitchen utensils and water drums. These artisans can be local, but are mainly Hausas from the north of Nigeria who move southwards during their off-farm season. Care of children, elders and domestic animals is the responsibility of the women, and women prepare the meals and do the laundry although they receive assistance from the children. Women's special duties revolve around the *agwus,* the troughs used for palm oil processing, where a high level of hygiene is required. Children sweep the rooms and the entire compound including the *atakpa (*the 'relaxation parlour'), under the supervision of their mothers. The maintenance of the *atakpa* is the exclusive right of the men.

A striking feature is that so many roles can be covered at the same time. It was observed in the course of this work that one woman transacted six duties at the same time. Her first job was cooking palm fruits. While the drum of oil was in the making she roasted yams and fed them to ten people. Cassava was also being peeled under her expert eye. Meanwhile she and her children collected cassava peels and used palm bunches and arranged them for compost. While this was in progress a trader with a plot of cassava for sale was welcomed to the compound. A deal was successfully concluded. In the household at this time were seven nieces and nephews whose mother had died. Still upset and lonely she gathered them under her 'wing' holding their hands and patting their heads as she went about her business. She also had a word for anyone who went by. Her satisfaction was that she could ensure happiness and contentment for those whose well being and livelihood relied on her. Male occupations do not lend themselves to such strengths but every man and woman returning home from the farm brings fire wood and grass for the animals.

Community maintenance is again gender related, though here the age grade groups play a dominant role. Young men from a particular age grade maintain the roads, community squares, schools and other community owned properties especially the palm and other economic trees. The female age grades ensure that halls and other buildings are swept daily.

The overall provision of adequate food is the responsibility of the household head. The wife ensures that it is ready at the time required, and provides the ingredients for the soup. At the end of the harvest the household head allocates stored food for consumption, and surplus, when available, is sold. Men eat in the *atakpa*, where their sons often join them; women and children eat by the kitchen. Breakfast is not normally prepared, but soup left over from the previous night is reheated for the children before going to school. If the household head is generous and rich enough, the family may enjoy *akara* and *akamu* before going to school or to work. The men go to farm without eating, but they will prepare a small meal there at mid morning.

The crops for household use are produced on the many plots that comprise the family farm, but in addition to this women have their own plots of land where the proceeds belong entirely to them. Sometimes the female 'managed' plots are used to produce ingredients for soups, but the produce may also be sold and the cash used for trading.

Women do the marketing, although men monopolise the sale of palm fruits, palm wine and meat. Women are allowed approximately 10 per cent of the income from sales of household produce. The household head allocates funds for school fees and all other contingencies such as food, clothes and hospital bills. Nowadays, the men welcome the initiatives of women, according to them.

One hand and head could not handle all the problems

The alignment of demarcation lines in cultural activities, with men producing vegetables and women owning tree crops, shows how interdependency has developed with a growing appreciation of the role of women in the well being of the village.

A discussion on social class in Eroke is difficult as there is little difference between lifestyles in the community, although certain inequalities exist. West Africa differs from India in that landlordism, renting and daily paid farm labouring is unknown. There are differential and transmittable rights over land in West Africa, but there is scarcely a household so poor as not to be able to farm. Rural class stratification is therefore not an issue and the equivalent of a class system does not exist. Ruling classes do exist in West Africa, but royalties bring little economic benefit. Women are excluded from the offices of *gagu* and *madaki*, but they can receive honorary titles with some jurisdiction over female activities.

While men still traditionally own and control tree crops, they sell part of the unprocessed palm fruits to the women. Indeed, one of the most interesting and contentious findings of this research for Igala living outside of Eroke was that women were now inheriting tree crops from both their fathers and husbands. This is a relatively recent trend that literally had to be seen to be believed by some as it was 'simply impossible in Igalaland'. In many Igala villages it is unheard of for women to inherit trees as they do not own the land. The change in Eroke has partly been brought about by the decline in the male adult population, but also by a desire of the fathers and husbands to ensure that their womenfolk have a reasonable source of income on property indisputably belonging to the family. Women are also increasingly involved in self-planting of trees on their husbands and fathers' land. Female ownership of trees in Eroke ends at their death, but this too could change given that women now take so much overall responsibility for the reproduction and maintenance of the human and natural resource base. Given that female ownership of trees was once regarded as 'impossible', it now looks as if anything is possible.

Religion and feasts

The three main religions (Islam, Christianity and traditional) co-exist peacefully in Eroke, as they do throughout Igalaland. All have their own places of worship, and the buildings are close to each other. Each day begins with prayer, and a gong (made from an old wheel rim and a piece of iron) summons household and compound members to their places of worship. Prayer, in whatever faith they profess, is an integral part of life in Eroke, and it is during communal prayers that one can become aware of concerns closest to their hearts. They pray for their children's future, household, farm and crops:

> sickness or any devil that would bring evil to their household be driven away.

> guide them in their health and business.

> that any evil or enemy causing harm to their farm and their crops be driven away.

> a gate be placed between the enemy and their farm.

> No prayer is complete without invocations for:

> the blessing of a male child.

Those with children 'abroad' always remember their immediate needs, especially deliverance from accidents and security of their employment.

Christians and Moslems still have ties with their traditional religious roots, and the whole village worships the land goddess (*Ane*). Other traditional religious practices such as *Achemuduje* (food prepared and eaten), and *Iyalo-Igala* (the end of year celebration) are common to the entire village.

There are a number of cultural feasts celebrated each year in Eroke. One, known as *Ote-Egwu*, is a form of ancestral worship specific to Imane district and occurs twice yearly. Everybody takes the preparations and celebrations seriously, and even the 'abroadians' welcome the excuse to return home and take part in the celebration as it enables them to maintain their links with their village. The whole community is united in prayer to thank their ancestors for giving them protection. At this feast they pray for all the intentions mentioned earlier as well as the progress of Imane district. Births, deaths and marriages are all celebrated, and death is celebrated as much as birth.

Leisure

People in Eroke have a capacity for 'enjoyment' equal to any other village in Nigeria or elsewhere. Regardless of what constraints or sanctions the country may be experiencing, the appetite for and ability to relax cannot be removed by any internal or external force. Leisure times are as sacred as any other activity, and an analysis of livelihoods in Eroke would be incomplete without leisure. Men and women rest after mid-day meals and share with each other the experiences of the morning. Children have their stories from school which are always a joy for the adults to hear. Everybody meets again in the evening often in the *atakpa*. People listen to their radios and discuss the current national and international political and social events. Recent romances and the prospects of a marriage always brighten up an evening. Weddings are a great occasion for relaxing, with young people often 'seeing themselves' for the first time in their 'best cloth'. However, the greatest time for relaxation is during the new moon with its ritual of entertainment and celebration and initiation of children into the culture of the community.

Fridays and Sundays are days where farm work is avoided as much as possible. Instead, Christians and Moslems devote their time to leisure and corporal works of mercy. Visits are made to the sick, bereaved and families with new born babies. These have an important social dimension showing how much care each community member has for the next. News and views are again shared, and information spreads more easily than if more formal structures were employed. People relax in each others compounds, palm wine often helping to dispel any tension or traces of animosity that may be there.

Indigeneous financial institutions

The indigenous, informal saving system is an age old institution playing an important role in the socio-cultural and economic life of people in West Africa and particularly in Igalaland (McNamara and Morse, 1998). Known as *oja* in Igalaland (*esusu* in Yorubaland), its significance in the daily life of households is understandable as capital accumulation is problematic. Nowadays the need for a lump sum of between N5,000 and N10,000 for school fees is common, and examination fees alone are over N3,000. Traditional *ojas* are still alive, but as an adaptable and flexible institution new forms are always being gradually introduced and accepted.

The benefits common to most *ojas* are cash contributions, financial assistance, payment of hospital bills, helping members with their farms, buying uniforms, funeral costs (coffins, white cloth, gun powder, grave preparation etc.) and visits to bereaved families or members with a new baby. Some of the activities are *oja* specific. All *oja* members have special celebrations at Christmas and Sallah. Being a member of one or more *ojas* simultaneously is common place, but can leave members without adequate cash flow. However, there is the assurance that these investments can be mortaged to a lender should there be any unforeseen calamities. Ability to absorb shocks is helped by 'joining local contributions'.

Given that the state does not provide children's allowances, old age or widows' pensions, disability support, health care or education, the *oja* is the only type of insurance policy available in the local context. In many ways, the *oja* instils corporate values in a community without government welfare services; the sick are visited, the bereaved consoled and as far as possible the hungry fed. The latter is possible through voluntary contributions from members, and/or 'bag' money, a fund that is allowed to accumulate overtime and derived from special levies paid on days when members receive their lump sum. Money realised from late comers, absentees and other penalties imposed on 'noise makers' and those failing to adequately discharge their assigned duties, are pooled for such purposes.

Many Eroke people invest much of their income in *ojas* that provide for their own death and funeral as these are very significant in the traditional religious and cultural beliefs although they do not live in the shadow of death. The quality of 'life after death' is determined by the quality of the funeral, hence the need to prepare adequately for it during one's life. Even with this foresight, which must force them to sacrifice some pleasures while alive, funerals are still a major expense for the family of the deceased.

There are a number of *ojas* in Eroke. Their present position will be discussed in Chapter 6 together with some recommendations as to how they can be more effective in the present dispensation.

Self-reliance in Eroke

The inhabitants of Eroke have always taken great pride in their history of self-reliance. What is equally important to them nowadays is that this spirit is still alive, but adapting to the many changes that continuously occur in their lives. The evolutions that have taken place and the forces driving them can all be clearly charted. The introduction of corrugated sheet roofing eliminated fire hazards for hitherto fire leapt from one thatched roof to the next. Eroke is now safer, though this clearly emphasised the difference between the better off and those not able to afford this innovation. It likewise meant less work for thatchers. Migration and education influenced traditional forms of self-reliance, but important indigenous knowledge was preserved. A good example of this is found in traditional medicine. Though western medicine was widely accepted many returned to the towns and cities with medicinal ingredients which they prepared for themselves as the need arose. Things became obsolete if they no longer had a practical use, but even to this day there is some knowledge of how things were done in the past. Making local salt is a good example of this.

The situation in Eroke in the past can be seen in many ways as a high-point in self-reliance, yet they were obviously quick to recognise what they saw as 'better things' and did not sentimentally cling to the 'old ways' when more appropriate options were available. The rate of change was, and still is, slow because they cannot afford to experiment with money required for urgent and immediate needs. They prepare and plan, and all decisions are carefully deliberated upon before any action is considered. Rarely if ever would rash decisions be taken as these would be a major threat to survival and sustainability. Certain decisions taken in the past, especially those relating to education, did not favour Eroke in the long-term. This situation is being rectified as every child of school going age is now in school. They have learnt from their mistakes, and are determined not to be left behind again. There are certain areas where new knowledge could again be beneficial. Fast growing trees for fuel wood is an immediate consideration, but it may be more beneficial to plant these when old oil palms are finished and more land is available on these oil palm sites.

This chapter introduced us to Eroke. Once acquainted with its history, geography, indigenous institutions, its forms of self reliance, its customs, and its leisure activities one gets the notion of a dynamic community adapting to change vital for its survival. The main actors in this drama will be introduced in the next chapter. Their commitment to the struggle for a better livelihood for their children will be highlighted as they work unrelentlessly to provide them with education which they believe will ensure a secure and sustained future.

Table 3.1 Number of Madakis and Gagos in the 'Abo' group of villages

Village (Abo)	Number of gagos	Number of madaki	Clan name	madaki area
Inele	1	2	Ijejo	
Ojoche		2	Iduh	
Eroke	1	1	Ujah Otekwu Igboyi	Eroke Ogane
		1	Ejenbi	Eroke Ate
		1	Ikoh	
Atanegoma	1	2		
Itekpe		2		
Ojuwo	1	2		
Ogbodo	1	1		
TOTAL	**5**	**14**		

The three villages, Abo Inele, Abo Ojoche and Abo Eroke are usually referred to as 'Eroke'. A *madaki* is the leader of a clan, while a clan can be thought of as a 'kinship group'. However, some clans can have more than one *madaki*, and smaller clans can unite under one *madaki*.

Table 3.2 Traditional councils in Kogi State, Nigeria

Traditional council	Notes
Kogi State Traditional Council of Chiefs	current Chair is the Attah of Igala
Area Traditional Councils (ATCs)	five in Kogi State corresponding to the five ethnic groups: Igala, Igbira, Yoruba, Bassa and Koto Karifi. The Igala ATC is responsible for the *onus* in Igalaland.
Local Government Traditional Councils (LGTCs)	nine in the Igala ATC corresponding to the nine local governments. These are responsible for the *gagos* in Igalaland.

Table 3.3 Population of the 'Abo' group of villages and Eroke

(a) Population of the 'Abo' group and Eroke

Category	Code	Age range	'Abo' group 1963 Official census	Eroke 1995 study census
Female elders	FE	>65	29	60
Male elders	ME		13	35
Female adults	FA	18-65	502	346
Male adults	MA		358	247
Female children	FC	<18	656	375
Male children	MC		596	322
Total			2154	1385

Notes The 1963 Official census is still more accepted than the census of 1973 and 1991. Eroke is usually assumed to comprise about 50% of the total population of the 'Abo' group of villages. Therefore in 1963 the Eroke population was approximately 1077, and the increase between 1963 and 1995 was 29%.

(b) Population in each of the seven Eroke clans (1995 study census).

Clans	Number of comp.	HH	Elders F	Elders M	Adults F	Adults M	Children F	Children M	Total
Ejenbi	9	14	4	2	43	38	46	36	169
Iduh	26	47	25	13	103	64	95	88	388
Igboyi	6	20	14	8	41	27	47	22	159
Ijejo	12	14	9	5	31	22	30	28	125
Ikoh	12	21	2	3	66	46	99	86	302
Otekwu	1	6	1	2	32	24	24	24	107
Ujah	6	13	5	2	30	26	34	38	135
Total	72	135	60	35	346	247	375	322	1385
			95		593		697		

Total number of compounds (comp.) in Eroke = 72
Total number of households (HH) in Eroke = 135
Total population of Eroke = 1385 (females = 781; males = 604)

Table 3.4 Household population and population change for a selected sample of 20 households in Eroke

(a) Household population in five 'anchor' years

Year	Children M	F	Adults M	F	Elders M	F	Totals M	F	Totals
70	50	36	46	51	4	7	100	94	**194**
80	59	48	58	62	3	4	120	114	**234**
90	60	42	63	73	4	4	127	119	**246**
95	55	47	50	61	6	8	111	116	**227**
98	57	57	45	53	5	6	107	116	**223**

(b) Household population change between the anchor years

Period	Births M	F	Immig. M	F	Deaths M	F	Emig. M	F	Balance M	F
70 to 80	30	23	3	14	9	12	4	5	+20	+20
80 to 90	30	24	5	10	9	6	19	23	+7	+5
90 to 98	22	30	13	22	19	9	36	46	-20	-3

(c) Summary of population change in the 3 age categories

Period	Children	Adults	Elders	**Balance**
70 to 80	+21	+23	-4	**+40**
80 to 90	-5	+16	+1	**+12**
90 to 98	+12	-38	+3	**-23**

Table 3.5 Some of the common crop species found in Eroke

Group	Botanical	Igala (English) name(s)
Root crops	*Dioscorea rotundata*	uchu (white yam)
	D. alata	ebina (water yam)
	D. cayenensis	ogoma (red yam)
	D. dumetorum	ulahi (bitter or trifoliate yam)
	Ipomoea batatis	uchu-opa, uchu-omu (sweet potato)
	Colocasia esculenta	ikachi (cocoyam or taro)
	Xanthosoma sagitifolia	ikachi (cocoyam or tannia)
	Manihot esculenta	abacha (cassava)
Cereals	*Zea mays*	akpa (maize)
	Sorghum bicolor	okili (sorghum)
	Pennisetum typhoidium	okadu (millet)
	Oryza sativa	iscapa (rice)
Legumes	*Vigna unguiculata*	egwa (cowpea)
	Arachis hypogea	opa (groundnut)
	Sphenostylis stenocarpus	yam bean
	Cajanus cajan	agwugwu (pigeon pea)
	Glycine max	soyabean
	Vigna (=Voandzea) subterranea	okpa (bambara nut)
Others ('soup' crops)	*Sesamum indicum*	igogo (benniseed)
	Citrullus vulgaris	api (egusi melon)
	Hibiscus esculentus	oro (okra)
	Lycopersicon lycopersicum	tomato
	Capsicum fruitescens	akpo (chilli pepper)
	Amaranthus spp.	alefo (amarathus or spinich)

Table 3.6 Some common economic and firewood trees in Eroke

(a) Economic trees

Botanical name	Igala (English) name(s)
Elaeis guineensis	ekpe (oil palm)
Cola nitida	obi (kola nut)
Citrus sinensis	alemu (citrus or orange)
Mangifera indica	umagolo (mango)
Prosopsis africana.	ukpehie (locust bean)
Parkia biglobosa	ugba (locust bean)
Borassus aethiopum	odo (date or fan palm)
Psidium guajava	guava (guava)
Irvinga wombulu	oro-ayikpele (oro tree; bush mango)
Irvinga gabonensis	egili (bush mango)
Crescentia cujete	oli-ugba (calabash tree)
Musa domestica	ogede (banana and plantain)
Anacardium occidentale	cashew
Carica papaya	echibakpa (pawpaw)

Table 3.6 continued

(b) Firewood trees

Botanical name	Igala (English) name(s)
Anthocleista nobilis	odologwu, odogwu
Daniella oliveri	agba
Vitex doniana	ejiji
Chromolaena odorata	abilewa (eupatorium)
Lophira lanceolata	okopi
Ficus carpensis	ugbakolo, ogbakolo
Morinda lucida	ogele
Melicia excelsa	uloko
Adansonia digitata	obobo
Combretam sp.	itado
Ficus iteophylla	oda
Azadirachta indica	oli-oda
Lecaniodiscus cupanoidies	okpu
Dialium guineensi	ayigele
Nauclea latifolia	ogbahi
Ceiba pentandra	agwu (cotton tree)
Cissus ciliata	ugbolo-eko
Stercubia tragacantha	abonoko
Newbouldia laevis	ogichi

Note Some of the 'economic' trees are also used for firewood, and some of the 'firewood' trees also have other products (fruits, bark etc.) used as food ingredients and also in local medicines.

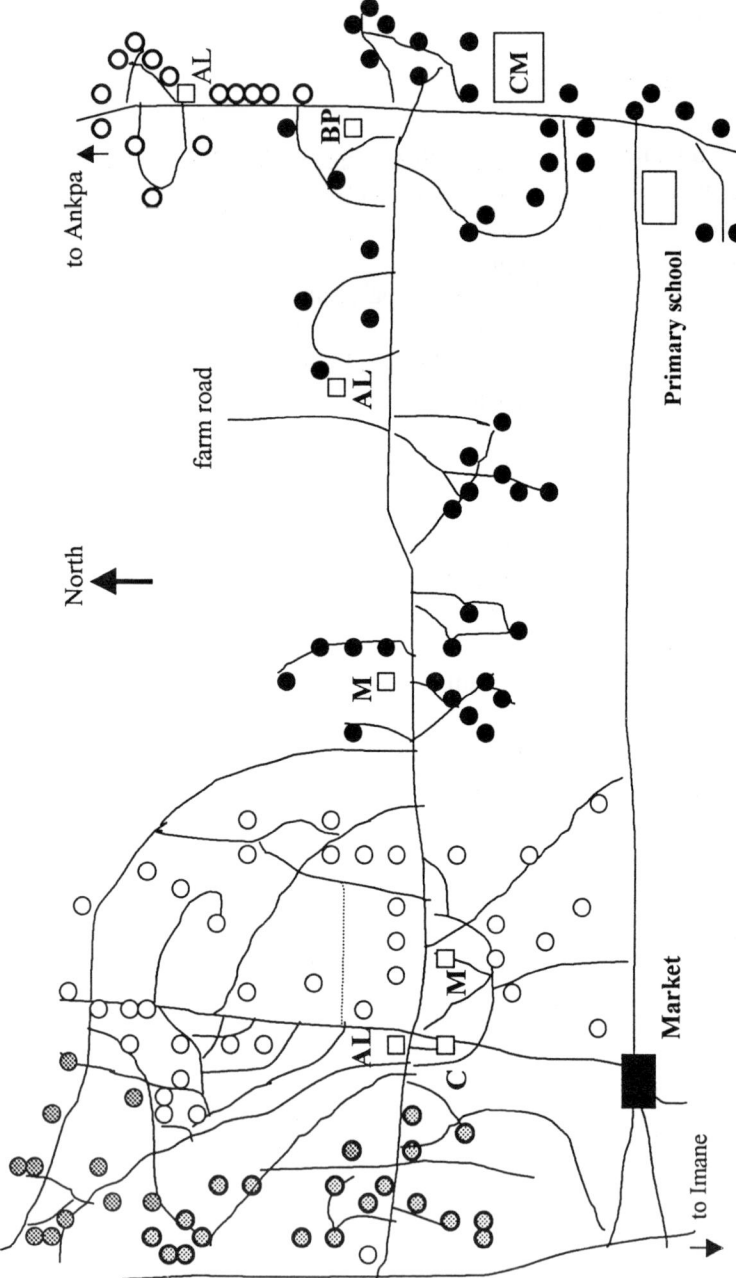

Figure 3.1 Map of Eroke village. Major roads and footpaths are shown along with the compounds

Key

Symbol	Clans	Madaki no.	Area	Village
○	Ujah Otekwu Igboyi	1	Eroke Ogane	Abo Eroke
◉	Ejenbi	1	Eroke Ate	
◉	Ikoh	1		
●	Iduh	2		Abo Ojoche
○	Ijejo	2		Abo Inele

C   church
M   mosque
CM   Catholic Mission
AL   *Alekwuanya* (women's traditional meeting place)
BP   business premises

The three villages Abo Eroke, Abo Ojoche and Abo Inele are collectively referred to as 'Eroke'.

4 Some Life Histories

Introduction

The previous chapter explored some of the novelty in Eroke. This chapter provides a further illustration of this by looking in detail at the lives of a number of people who live there. The aim is to reinforce some of the key points raised in Chapter 3, namely the importance of education and the fact that Eroke is certainly not an island set apart from the rest of Igalaland, Nigeria or even West Africa. These two points are related in that education has been, and continues to be the passport to employment outside Eroke.

The second objective is to stress the obvious fact that sustainability is about human beings– real people, with aspirations and expectations – and not numbers or statistics. Just as 'environment' is a human construct (Blaikie, 1995) then without individuals, groups and communities there would be no concept of sustainability. As the sustainability literature puts emphasis on quantification of systems, scales, flows, production and quantities, it is in danger of moving into quantitative abstraction, and so in this chapter women and men are viewed in real life situations. Even 'social' aspects of sustainability, in so far as they are discussed at all, are typically consigned to a few abstract notions such as 'social welfare', 'sociological benefits' and 'quality of life' often appearing at the bottom of SI lists almost as an afterthought or a gesture. This results in critical areas being subsumed into a few quantifiable measures such as income and 'income distribution'. This chapter seeks to avoid this trap and the presentation of detailed life histories midway in this book will help focus on those who live in Eroke.

Those selected come from a variety of backgrounds. Most are indigenes of Eroke, and though not necessarily full-time farmers would engage in it in some capacity. The authors had to be selective when choosing the storytellers, and endeavored to include the wide range of skills and experiences available there and in most other villages in Igalaland. It is important however, to stress that everyone in Eroke has a story worth telling and are more than willing to talk about their lives and aspirations. Those who gave their histories were simply asked to speak freely about the most significant events in their lives. Because they all wished to speak about their children, especially regarding education, this information is included in Table 4.1. Readers are now introduced to actors important in the daily life of Eroke:

- The Okoliko trio, Lazarus, Jerome and Matthias (brothers).
- The Ohemu brothers (blacksmiths): Hassan and Zekeri
- John Ochimana (farmer and teacher)
- Francis Ameh (driver)
- Joseph Ejeh (farmer)
- Abdullahi Adama Akpihi (herbalist)
- Ali Shaibu (entrepaneur)
- Emma Abah (*madaki*)
- Asana Idoka (female herbalist)
- Helen Akpihi (trader)

The Okolikos, John Ochimanna and Joseph Ejeh are full-time farmers and representative of most farmers whose wives also trade. The Ohemu brothers, Francis Ameh and Abdullahi Adama Akpihi are specialists, and only farm on a part-time basis. They represent a sizeable minority of males for whom agriculture is important, but not the mainstay of their livelihood. Ali Shaibu, a young entrepreneur, is by Eroke standards, rich. He represents the upper extreme in Eroke, but one to which many would aspire. The two women in the sample are quite different. Asana Idoka is a herbalist, but unlike Abdullahi she was neither trained nor initiated into this profession by her father. Hers is a moving story as she had no ambition to enter the healing profession. Helen Akpihi is possibly most representative of Eroke women in so far as she is a trader. Indeed Helen was the driving force motivating many women traders in Eroke. She pioneered long-distance trading in the village, and has built a chain of contacts extending to the north and south of Nigeria.

The Okolikos (farmers): Lazarus, Jerome and Mathias

These three brothers belong to the Igboyi clan, and share the same compound. Lazarus is the compound head, but all have independent households. They each farm on communal lands and utilise the more fertile land surrounding their compound. Farming of communal land is free, except for the tithes payable on economic trees. Their father, born about 100 years ago, had only one wife which was unusual for that time. Six children were born to this family, but the eldest brother died leaving Lazarus to assume family responsibilities at a very early age. To do this he became a farmer, petty trader and hunter simultaneously. He never went to school, and never had anyone to lend a helping hand. When it came to paying dowry for his first wife he had to fend for himself. When married he moved to a separate household with his new wife, leaving his late father's house for the younger members of the family. All the brothers co-operated in the construction of the new house. Lazarus and his wife had five children, but three died at an early age. The remaining two children have done well. The boy is a civil engineer and the girl, though not educated, is 'well married'. He is satisfied with their progress and feels they have done better than himself, something that gives him contentment.

Lazarus divorced his first wife, and remarried. He had seven children by the second marriage, but again tragedy struck as four of the children have died. The surviving boy has completed secondary education, and gone to live with a cousin in Ilorin (Western Nigeria). There he hopes to receive third level education as this relative has promised to pay the bill. The two girls are also in school, and as far as he can see 'opportunities have been put in their path'. The well being of his children is more or less secure, 'in so far as it lay within my power'.

Lazarus became a Christian fifteen years ago and claims this brought him peace of mind as he found some aspects of traditional belief disturbing. He sees land as a potential source of conflict in Eroke. Should any of his two sons, each from different mothers, return home there could be problems if both required land. Lazarus' wife is a farmer, producing mostly cowpea and cassava. This is of great benefit as he is now in poor health for the first time in his life. He has to spend much money on medical treatment. He has never traveled much outside Eroke except when he traded. He has had no opportunity to accumulate any capital for retirement, but hopes his sons will be generous to him. Lazarus feels that facilities such as a hospital and nursery school would help entice people to return to Eroke, but the poor soil of the area is a major problem.

Jerome Okoliko is also a farmer. He did not go to school. His parents died when he was young but he can still remember them. Lazarus took over their role and trained him on the family farm where they worked to provide food. He was always glad to see Lazarus when he returned from trading. Eventually he saw a girl he liked and Lazarus paid the dowry for him. Two of the four children from this marriage died, and the trauma surrounding their deaths ended in a divorce. The daughter, who went to primary school, is now married and the son, who completed secondary education, works in Kano for a textile company that is sponsoring his third level education. Their success is a source of some satisfaction for him. They have the 'things I never enjoyed and happy because they have them in my own lifetime'.

Jerome remarried, this time to a widow from a nearby village. She and Jerome have two girls, one is in nursery school and the other in primary school. His wife had two girls and a boy by her previous marriage, the boy is in secondary school in his late father's village. One of the girls is married and the other is still with Jerome and her mother. Understanding what it is like to be an orphan motivates him to ensure his step-daughter has what she needs for a secure future.

In 1958, at the age of sixteen, Jerome traveled with his friend and neighbour, John Ochimanna, to Ijebu Ode in the West of Nigeria. Here they worked as labourers for a year. While there, they learned of good employment opportunities in Equatorial Guinea, a Spanish colony at that time (Figure 4.1). He and John returned to Eroke and after a few months traveled to Enugu and Calabar in South-east Nigeria to pursue the work prospects in Equatorial Guinea. At that time (1959), the British colonial government in Nigeria had an agreement with the Spanish colonial government of Equatorial Guinea regarding staffing of plantations, and recruitment took place in the south east of Nigeria. Here they met other Igalas who provided advice and helped them obtain passports. In due course, they signed a contract with the plantation company and traveled by ship, free of charge, to their new destination.

This new experience was somewhat daunting for Jerome, particularly as he had no English. Directions for work were given first in Spanish, then translated into English and finally interpreted by Igalas working on the sites. While in Guinea they received a local allowance, food and lodging. Workers could farm on land allocated to them, and were also allowed to hunt and cut palm bunches. As work began at 7.00am and finished at 3.30pm it was possible to earn extra pesetas from the plots allocated to them. Changing pesetas to pounds was problematic, and it was forbidden to bring back any cash except the salary paid into a bank account in Calabar. This hurdle was overcome by sewing the pounds neatly into hems of trousers and pressing them so well that no one noticed. Others, especially couples smuggled back their surplus concealed in foodstuffs. Jerome and John spent three and a half years in Equatorial Guinea, first on a two-year contract that was then renewed for a further one and a half years. Jerome was struck by a number of things in Equatorial Guinea. One that touched him greatly was 'that white and black were suffering together on the plantations'. He did not think this was possible.

They returned to Nigeria in 1962 and collected their salaries in Calabar without any problem. On returning to Eroke Jerome worked as a laborer, mainly as part of the *ogwu* system, in nearby villages as well as farming for himself mostly at weekends. In 1976 he became seriously sick. This was a big set back as this particular ailment did not respond to either western or local medications, and to this day he has not fully recovered. His first wife found the stress of this combined with the deaths of two of their children too much to take and she divorced him. Jerome eventually remarried. His new wife is a farmer and trader and she has certainly been helpful to him. He feels he worked hard all his life, but the greatest enemy to progress has been 'depleted land and poor yields'. He likes Eroke, but would prefer his children to enjoy a life 'more free of hardship' than he has had. Like Lazarus he thinks Eroke needs a nursery school and a hospital, as well as a good road to entice the youth to stay there.

Mathias is the youngest of the three brothers, and has no recollection of his parents. He was brought up by Lazarus and didn't attend school. Like Jerome, he went at an early age to the West of Nigeria as a migrant laborer on the cocoa and yam plantations. He came back to farm for three years, but returned to the west in order to earn some money. Lazarus invited him to return, built a house for him in the compound and offered to pay the dowry for his wife. Mathias had already chosen the girl, and Lazarus did the big brother 'thing' by facilitating the arrangements. Senior brothers were expected to assist in such ways.

As Mathias' wife was previously married, she came with her daughter. They now have two daughters and a son of their own, all of whom are in primary school. His stepdaughter is now engaged. His wife, though still young, hasn't formal education and is a successful trader and farmer. This is fortuitous though Mathias still finds life difficult as he continuously struggles with a small income from poor land from which he has to pay education and health bills. He hopes his children will have an easier life than he has had, and believes this is likely to be outside Eroke.

The Ohemu Brothers: Hassan and Zekeri (blacksmiths)

These two brothers (Hassan and Zekeri) are blacksmiths and farmers. They live in the same compound, but have different households. Their father, who came from Ajobe, a village 16km away, married from Eroke. At that time his father-in-law invited him to settle in Eroke. He subsequently married six more wives none of whom were from Eroke. Three of these later divorced him because they had no sons. Another wife was barren and lived with him till death. Their own mother bore seven children.

Their father was a blacksmith and farmer, and both had learned blacksmithing skills by the time he died. Hassan has two wives, one from Ankpa and another from Imane. The first wife has a son and daughter born around the mid-1940s. The daughter didn't attend primary school, but the son who did became a member of the mobile police. He was in active service at the time of the Nigerian civil war (1966-1970), and now lives in Kano. His second wife, who is much younger, has ten children.

Hassan's first wife, though now quite advanced in years, is a salt moulder. His second wife is a local trader, buying processed palm oil and selling in nearby markets. Training children was a joint effort between the wives and himself. It is now twenty years since he became a Muslim, and like Lazarus Okoliko he recalls some aspects of the traditional religion which were disturbing. A friend introduced him to Islam, and now his own family as well as the extended family have embraced the Muslim faith. His eldest son by the second wife is in Lagos, and was a Christian before the rest of the family converted to Islam.

Hassan farms on communal land in Eroke, so he only pays tribute on the palm trees. However, most of the household income is derived from the blacksmith business. He and his brother run the business jointly, producing hoes, axes, cutlasses, local traps, 'Dane' guns, the special axe for harvesting palm trees, the chisel for palm wine tapping and special knives required for peeling cassava and yam tubers. They also repair implements. They market their products in Eroke and neighboring villages. All the male children are equally competent in the blacksmith business. They offered training to other indigenes of Eroke, but without much success as only one completed a full apprenticeship. Hassan is now close to eighty years of age. All his children have left Eroke except for some daughters still in primary school. He prefers to stay there. He occasionally visits his sons and brothers, and traveled to Kano for an eye operation some fifteen years ago. He is thankful that this was successful. His children paid all expenses.

Zekeri also married twice. His first wife has eight children, while the second has three (one son died). Zekeri himself never traveled though his second wife is a long distance trader – mostly to Kano. She sells oil palm and oranges and returns with cowpea, millet, locust bean and benniseed which she finds in markets in Tiv and Idomaland. His first wife is a farmer and trades part-time in local bean cakes which she produces herself.

Both these men felt there are only limited opportunities in Eroke, and according to them it is almost sure that the only young man in the family who is interested in farming will return to the more fertile soils of Ajobe, where he is entitled to clan land.

John Ochimana (farmer and teacher; born 1936)

John is popularly known as 'Teacher'. His father had two wives, and he 'belonged' to the first wife who had three sons and two daughters. His father's second wife had only one son. John resides in the compound of his maternal grandfather, in line with the Igala tradition which allows a man to make his home, in that of his father, mother or wife. His father died when he was eight years old. He and his mother then left his father's compound and came to stay with his uncle in Eroke. After some time with his uncle, he was sent to his elder brother's friend. However, this gentleman refused to send him to school so he returned to Eroke in 1949. He finally managed to attend primary school in Imane – fulfilling his life long dream. In 1956 he became a Christian Religious Instructor (CRI teacher), and from this occupation came the title 'teacher'. He worked in villages close to Eroke for two years. His meager salary was partly paid by the Catholic Mission and the villagers. This system was an effort at sustainability in pastoral activities, but was not very financially satisfying for John.

In 1959, John traveled with his friend, Jerome Okoliko, to Equatorial Guinea to work as a labourer on a plantation. Since John was literate he was given a supervisory post which involved the control of fifty labourers. Supervisors enjoyed better conditions than labourers, including a better salary, more food (double the ration of labourers) and better accommodation. John enjoyed his time in Equatorial Guinea, and while there became aware of the importance of economic trees.

Upon his return to Nigeria, John went to Jos as a labourer in the tin mines. In 1966 he married a woman from Ogoja (a sister of his co-worker), and they had three children. They stayed in Jos until 1971 when once again he returned to Eroke to take up farming and assume responsibility as household head. In 1974 he was given a post as an agricultural extension agent with DDS and the World Bank project in Igalaland. He attended agricultural college in 1980. He married a second wife (Helen) in 1979, but they have no children of their own. His wife however came with her two daughters from a previous marriage, and John educated them to primary school level. One trained as a tailor and both are married. Helen is a successful local trader, and has learnt much about farming from her husband. He eventually took a third wife who had a daughter. Later they divorced, but the child is in primary school in Lagos staying with John's son.

John has much land and many economic trees, mainly oil palms, coffee and citrus. These he bought from the Ministry of Agriculture when he returned home in 1962 and again in 1971. All his children have left Eroke, but his grandchildren are now living with him in the household. His brother, who lived next to John, recently died but his sons did not return to assume family responsibilities. His brother's widow now heads that household, and many of her grandchildren live there while attending school. This experience provided John with much food for thought, who wonders if his own surviving son will be any different. It would seem that in two decades from now every household in John's compound will be female headed, and some may even have disappeared. This situation is very similar to that of many other compounds in Eroke.

Francis Ameh (driver)

Francis' father had only one wife who bore eleven children, although two died. Two of his brothers farm in Eroke, and two who had primary education work in Kano and Abuja. Two other brothers who went to secondary school, work in Imane and Kano. None of his sisters were educated, and his eldest sister is now dead. Francis was the fifth child, and completed primary education.

After primary school he went to Kano to learn electronics, but did not complete his course due to illness. He came home for treatment, but on his return to Kano the sickness re-occurred. His parents persuaded him to pursue a career in driving instead, and on the completion of his training he returned to Kano to work there in this capacity. However, he came back to Eroke after three months, and worked with the Imane Community Grammar School from 1978 to 1985. He married a wife from a village close to Imane while he was working at the grammar school. In late 1986, he was again employed as a driver by the catholic mission in Imane where he worked for three months. Throughout his employment he continued farming and combined this with palm wine tapping. He had another opportunity to work as a driver with the Federal Government College in Ugwolawo (a town in the west of Igalaland), but because he did not like the working conditions he resigned. In 1993, he was employed by the Kogi Agricultural Development Project as a driver and still works there.

Francis has two wives, and each has three children. Both wives farm and trade. He became a Christian in 1963, and was 'converted' while in primary school. He likes Eroke and has no regrets about staying there. His main ambition in life is to 'train his children very well'. He is sure that 'a good foundation' will give them security where ever they are; though Eroke may not provide enough opportunity for educated young people.

Joseph Ejeh (farmer)

Joseph's father was a chief in Ubele. He married from the Ejenbi clan in Eroke and had two wives. Joseph's mother was the senior wife, and gave birth to two girls and three boys, all of whom survived. She remarried from the same family on the death of his father, and had two more children. Joseph has never been to school, but his brothers attended primary for a short while. He was very young when his father died and does not remember him. When his mother remarried he and his brothers went to live with his maternal uncle who was a traditionalist. On the death of his uncle he returned to his father's compound and began to farm. He also 'trained' his younger brother and sister. Joseph has never left Eroke. He has one wife and five children (another son died). All his sons attended secondary school. His first son is self-employed in Osun State (Western Nigeria), and is married with four children. The second and third sons, who are already married, are in search of employment in Lagos. The fourth son is still in secondary school. His only daughter is married with two sons.

Joseph is a full time farmer on his mother's land in Eroke, where his father also worked prior to his death. He combines this with palm wine tapping. His palm trees are on community land so he 'pays tribute on them'.

His wife farms, but does not trade. He and all his family are Christian. He has no desire to leave Eroke, unless offered a chieftaincy title at Ubele. His interest in Eroke is largely determined by his access to palm trees and availability of water. He would be very happy if his children returned home, but thinks it is unlikely. He would never force them to do so.

Abdullahi Adama Akpihi (a herbalist)

Abdullahi is a full time herbalist, farmer and hunter. He considers healing his primary occupation. He inherited his herbal skills from his fore-fathers who were renowned for healing. Abdullahi has traveled and worked throughout Igalaland and Nigeria in order to gain skills and expertise in herbal treatment. He first began to practice in 1961, and attended courses in Northern Nigeria, Calabar and Lokoja where accreditation was possible.

The materials required for his work are all available locally. He often gives his services free of charge although people give gifts of cash or kind. In one case the gift consisted of the registration fee for his son for the West African School Certificate (a secondary level qualification). His satisfaction is in seeing people recover, and no one has come 'to report a negative aspect of his work'. He has a store in Ankpa where he dispenses treatment on market days, and many also come to Eroke for consultation. Some patients come from Cameroon for bone setting. Abdullahi is very committed to his work and passing on the skills to his sons. He sees this as a contribution to the community, and feels these will be a good supplement to whatever work they do, be it at home or abroad. He does not believe Eroke will stay 'the same for ever, it has even changed in the last ten years'.

Ali Shaibu (an entrepreneur)

Ali's father married only one wife, and the couple had ten children, including three sets of twins all of whom are still alive. Ali himself is a twin. The first son died, but four women and five men are still alive. Ali is now the elder of the family since his father died in 1997. His mother is still alive. Ali has no idea when he was born, and never went to school. His brothers and sisters went to school, and all progressed to third level education at polytechnics and universities. All of them are married and live outside Igalaland, but his brothers and sisters have houses in Eroke and return when they like. His father completely catered for the education of the male children, but Ali saw to the females.

Ali himself is a Christian though not yet baptised. When still young, and much against the wishes of his father, he went to Ado Ekiti, Uche Ekiti and Ilorin (all in Western Nigeria), where he worked on plantations. He lived in Yoruba land for fifteen years, and is now a fluent Yoruba speaker. While in Ado Ekiti and Uche Ekiti he was apprenticed as a farmer, and eventually became a master farmer. In Ilorin he rented land for farming. To help him in his work he brought young boys from Eroke and other villages in the Ankpa area to work as apprentices. After a year he paid them, took them home and returned with others. While in Ilorin he married an Eroke woman, built a house, bought a motor cycle, a mill for grinding grains, a bus and a car. He used the vehicles as taxis, but later sold them. He returned to Eroke only because his father was getting old.

Ali has four surviving children from his first marriage in Ilorin (one girl, died). He now lives with his second wife (Christiana) in Eroke and they have four children. A third wife died leaving him with a boy and a girl. Ali has other wives who live outside Eroke, some of whom have borne him children. He wishes to marry more. His declared reason for wanting many wives is that women do not like to have too many children nowadays, and in the event of an increase in infant mortality it is necessary to have many wives but each having only a few children. Therefore, his primary ambition is to have many children, especially sons, rather than many wives.

The Ujah clan, to which Ali belongs, made a decision to divide its land between all the families in the clan, and as a result Ali has a large area of land, possibly in the region of 40ha. His brothers could ask for some land at any time, and for this reason he is trying to obtain a Certificate of Occupancy. However, given the large area of land he owns he could easily 'give out land' to his brothers and still have plenty for himself. His ambition is to own a taxi and grinding mill as he did in Ilorin. Both he and Christiana are involved in trading.

Ali is an example of a highly successful, capable and ambitious entrepreneur who made full use of his resources and talents. He has good land and knows how to use it. He is quite ambivalent about the future of Eroke. At one level he is interested in Eroke and believes that it will continue as it is; the males going out initially but some returning later. He, however, acknowledges that returning poor to Eroke is problematic, and it is possible to have a better life style elsewhere. He would like his children to have as many options as possible in life, and believes that is what education has to offer. He certainly enjoyed his fifteen years in the west, and he could still combine work there with his responsibilities in Eroke.

Emma Abah (Madaki; born 1920)

The name 'Emma' means 'a child born for death to come and carry'. His father and mother had many children, all of whom died with the exception of Emma. Because of the death of the previous children, Emma was taken as soon as he was born in Eroke to an uncle in nearby Abo Inele, so 'that the spirits could not get him'. He returned to Abo Eroke as a mature adult, and has lived there ever since.

Emma married four wives, all of whom are now dead. He had nine children, but five of his sons have died. One of his sons completed primary education, and he farms in Eroke as well as in a nearby village. The other surviving son has completed his secondary education, and is in Lagos in search of work. One of his daughters went to primary school while the other has no formal education. Both are married and live outside Eroke.

Emma was once a full time farmer, but his son has that farm now. He is still an oracle consultor and has skills as a herbalist and bone setter (sets broken bones and fixes dislocated joints). He doesn't charge relatives or Eroke people, but outsiders have to pay. Emma is also a story teller of great renown. Each story has a moral and often takes the form of a rhyme, although he claims to have forgotten many. He is still a traditionalist, but all his children and grandchildren are Christian.

Emma is one of the *madakis* in Eroke, a position that helps supplement his income. He obtained this post between 1960 and 1965, and although he should receive some payment from government he has received nothing for the past ten years. However, he is exempt from paying tribute to landowners, and entitled to some of the communal palm produce.

He is of the opinion that Eroke will go on as it is, as the population has increased and western education has opened up many opportunities. His own son will succeed him, he already has his own household and he too is 'already old enough'.

Asana Idoka (female herbalist; born 1950)

Asana is a native of Adeh, a village 4km from Eroke, and has twice been a widow. She has a son and daughter from her first marriage. This son joined the Air Force in Port Harcourt after completing his secondary education. Her daughter died leaving five children; three of whom have since died and the remaining two are living with her and attending primary school. She was the third and last wife of her last husband, and since his death has been the household head. Asana gave birth to five daughters with her last husband, four are in secondary school and one in primary. She previously traded in kola, but is now a full time herbalist (*ogwu-erah*) and farmer.

The story of her healing began when her fifth child by her second marriage became seriously ill while she was still being nursed. She had tried local and western medicine, but none worked. She lay on her bed wondering and worrying about what would happen to her baby. In her sheer exhaustion she dozed and had a dream in which her late father appeared. He was in the company of many people none of whom she recognised. They inquired about the reasons for her worries and finally asked her to indicate 'whether her preference was for either her children or her goats'. 'My children' she replied, and they left. The next morning the goats began to fall sick and die, and she shared the goat meat with others in her compound. She didn't eat it herself. The following night the same group re-appeared in her dream, and led her to the bush where they showed her different kinds of herbs. They demonstrated on how to prepare them as cures for various ailments. Finally they asked her to give them a ram, and as soon as day broke, she sacrificed a ram. She immediately treated her own baby, using the directions received in the dream. The girl quickly recovered.

Asana continues to successfully prepare the various remedies, still faithful to the original instructions. All this was very new to her, although her father was a herbalist and consulted oracles. As a girl she wasn't taught any of these things by her father, and had no ambition to do anything in that line.

Asana is a traditional believer and so were both her husbands. Her children, however, are Christians, and she hopes they will return and build houses in Eroke and help her in old age.

Helen Akpihi (trader; born 1956)

Helen has been recently widowed (1997). This was unexpected. Her father had one wife and twelve children, but of these six died young. Both her parents are still alive. Her father came from a nearby village, and Helen was born in Eroke but never attended school. At the age of five, a female cousin took her to Ejule, the biggest market town in Igala. Here she became acquainted with trading as her relatives were involved in the marketing of red oil and grains. On the death of her cousin, Helen now eleven returned to Eroke and continued her career in trading. She stayed with her parents for a year, and was then betrothed to Abraham Akpihi. She moved to his house as soon as she reached adolescence. Like her own mother, she gave birth to twelve children, two of whom died.

Helen's trading continued to expand after her marriage, and she moved into long-distance trading to the north of Nigeria, particularly Kano, Bauchi and Jos, and southwards to Port Harcourt. Palm oil was in demand in the north, and she bought back dried fish, locust bean, cowpea and guinea corn for sale in Eroke and Imane. Some of her customers paid cash, but she also sold on credit. She never had problems in recovering her money with interest, although she tends to be very selective about who she lends to. She worked on the principle of quick turnover with small gain. She provided some goods to members of her own family so that they could sell in the local market and gradually gain experience in trading.

Changes in the Nigerian economy have had a major influence on Helen's business. In the time of President Shagari (1979-1983) she bought a bag of grain for N1,500 and sold it for N1,600; a gross profit of N100 (7 per cent of outlay). Oranges were more lucrative. Between 1979 and 1983 it was possible to load a lorry of oranges for N1,000 in Eroke, and when sold in Kano this generated a gross profit of N400 (40 per cent of outlay). Of course the cost of transport had to be deducted from the gross profit, but at that time the transport was relatively inexpensive and she could hire a lorry to carry her own goods. From 1996 onwards, as the cost of transport increased, she was only able to hire a lorry if she joined in a cooperative of eleven others (all from Eroke and Imane). A lorry load of oranges cost N12,000 in 1996, and yielded a gross profit of N4,000 (33 per cent of outlay), but she says that you could do far more with N400 in 1979-83 than you could do with N4,000 in 1996. As the cost of transport increased she concentrated on selling palm oil, gari, oranges and locust bean and buying tomatoes and onions as well as grains. She buys cloth for her family in Kano.

Helen, is always seeking alternative openings in the market, and recently she successfully acquired a new outlet in Port Harcourt. She has also moved into large scale processing of cassava tubers into gari, especially the new varieties that she purchases throughout Igalaland. She does this at her compound in Eroke. The traders cooperative, of which she is a member, has established its own *oja* meeting where they exchange market and contact information.

She has greater responsibilities since the death of her husband. They had always 'worked as real partners all their life'. She concentrated on trading, he on farming but both helped each other. Since his death life has been much harder, and this was especially so during the mourning period when Igala culture prevented her from engaging in trade. Helen believes it is only after their retirement ('when they resign') that her children will return to Eroke, but she would not like them to return yet. She is wondering if the population of Eroke will decrease as she sees that 'people are not coming back as before'. Many improvements have taken place in her lifetime, and she takes great pride in the fact that she introduced long distance trading into the Imane metropolitan area.

Life histories and sustainability

The life histories narrated here provide some useful pointers for an analysis of sustainability. First there is the obvious diversity of occupation and income. Although all who spoke used the term 'farmer' to describe themselves, this was not employed in an exclusive sense. Indeed some were more eager to discuss other occupations such as healing and trading than farming. Even those who considered themselves to be primarily farmers (the Okolikos, John Ochimanna, Ali Shaibu etc.) stressed the importance of other sources of income generated by their wives and children. One was left with the sense that to talk solely of agricultural sustainability would impose an artificial constraint on the conversation. Even so, agriculture is important and the next chapter will look in greater depth at issues of production, the resource base and consumption.

Secondly one was aware of the extensive networks that Eroke has with the outside world. Part of this was the experience gained by some in working elsewhere, the most notable examples being Jerome Okoliko, John Ochimanna and Ali Shaibu, but also the contacts that form an essential part of marketing. Helen Akpihi and her involvement in a traders cooperative provides a good example of the latter. Eroke is certainly not isolated from the rest of the world, and the importance of the 'abroadians' in supporting the households was constantly reiterated. All thirteen respondents loved to talk about their children, especially the ones who were living away from home. While Eroke has a clear geographical boundary, the perspective of many was bounded more by social concerns centered on the extended family, where ever they may be.

Thirdly, there was the obvious emphasis on education. All the households described in this chapter were managing to survive, and some were living well. However, on the other hand the fate of some of the households would not be very encouraging were it not for the young industrious women. The women trade, farm and earn many forms of supplementary income, and are single minded in their determination to send their children to school. Ironically the emphasis on education could be interpreted as sowing the seed for the future unsustainability of the village. Most believe that the educated will only find job satisfaction and job opportunity outside Eroke. Eleven of the thirteen who spoke about their children's lives saw very little future for them anywhere except in the cities. Some would like them to return 'but not yet'. They expressed hope that they would be 'good or generous to them in their old age'. Implied in many of their remarks was the desire not to be a burden.

The stark realities of emigration were all too obvious. Some households were disappearing, and grandchildren lived with their aged grandparents. These youngsters were sent there to give them a helping hand and/or to be educated in safe surroundings. As soon as the grand parents died, the children went back to their own parents and the household ceased to exist – at least in Eroke. The senior sons choose the city life, and there is no one else willing to risk returning home. The death of John Ochimanna's brother provides the best example of this out of the thirteen life histories, and certainly gave John plenty to think about with regard to his own son. The attraction of the city was plain to all, and no one was under any illusions. The cities provide more perks such as supplementary income, electricity and greater proximity to employment and market opportunities. Emigration of the young would at first appear to strike at the heart of agricultural sustainability in Eroke, yet this can also act as a safety valve and prevent excessive pressure on resources. Ali Shaibu provides a good example of this. Although he does own much land, if all his sons, brothers, and not to mention their sons returned demanding land then there could be serious problems. Maybe one future for Eroke is a smaller number of better-off people who can afford the amenities and who like living in the security of 'a safe village'.

Table 4.1 Some details for the thirteen families described in the life histories

Household head (* = F)	No. of spouses	No. of children live M	F	dead	Level of education P	S	T	N
Okoliko, L.	1	2	3	7	1	1	2	1F
Okoliko, J.	1	2	5	2	3	1	1	2F
Okoliko, M.	1	1	3	0	4	0	0	0
Ohemu, H.	2	5	7	0	7	3	1	1F
Ohemu, Z.	2	4	7	1	5	6	0	0
Ochimana, J.	1	1	4	1	4	1	0	0
Ameh, F.	2	3	3	0	1	5	0	0
Ejeh, J.	1	4	1	1	1	4	0	0
Akpehi, A.	2	6	6	0	0	12	0	0
Shaibu, A.	4	10	8	1	18	0	0	0
Abah, E.	0	2	2	5	2	1	0	1F
Idoka, A. *	0	2	6	1	3	4	1	0
Akpihi, H.*	0	8	2	2	1	4	5	0
Totals		50	57	21	50	42	10	5F
		107			107			
%		47	53	16	47	39	9	5

The number of wives/husbands refers to those currently living with the household head. The level of education refers to the highest level achieved as of the present. It is likely that many going to primary school will move on to secondary school.

P primary level education
S secondary level education
T tertiary level education (universities, polytechnics, technical schools and training for the armed forces and police)
N none of the above (illiterate). These were born in the 1940s long before it became fashionable for girls to go to school.

Some Life Histories 133

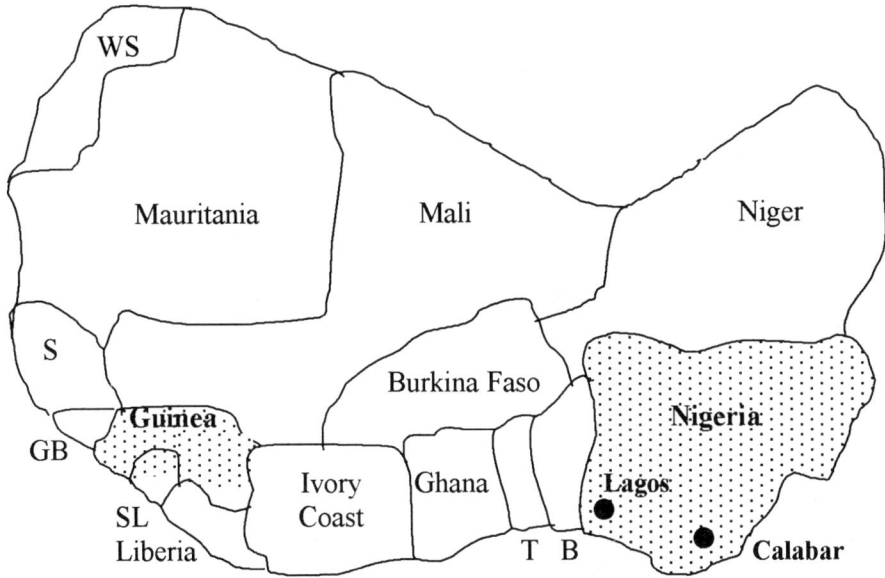

WS	West Sahara
S	Senegal
GB	Guinea-Bissau
SL	Sierra Leone
T	Togo
B	Benin Republic

The map is not drawn to scale (north is the top of the page).

Equatorial Guinea is where John Ochimanna and Jerome Okoliko worked for three years as plantation labourers.

Figure 4.1 Sketch map of West Africa showing the relative locations of Nigeria and Equatorial Guinea

5 Production and Consumption

Introduction

As discussed in Chapter 1, issues surrounding production are often central to many discussions on sustainability. Production is easily quantified, and from a western perspective yield or some monetary based derivative largely dependent on yield and its value, has long been a central consideration in agriculture. For many, sustainability is no more than the maintenance or increase in production over time, and a glance at the example lists of SIs in Chapter 1 quickly reminds the reader of the perceived importance of production. However, the form in which it is expressed does vary. Typical suggestions are SIs based on output/area, output/area/time or output/household, but monetary value and profit of production regularly appear. Others suggest (e.g. Isac and Swift, 1994) that appropriate SI's should be based on production as a proportion of what is achievable in the circumstances i.e. potential production.

Because of the importance attached by a western mentality to production, the resource base that underpins it is also critical to sustainability. Perhaps the most emphasised resource is soil as practically all lists of agricultural SIs describe its physical, chemical and biological properties. Soil erosion captures many imaginations as the penultum of unsustainability.

As outlined previously, agriculture is the mainstay of Eroke's economy, and in keeping with the rationale set out in the earlier chapters, this chapter will first focus on the complex issues surrounding production there and changes over the last ten years. It will begin with a brief outline of the cropping systems commonly found in Eroke, followed by a discussion on how these and production have changed over this period. Included within this is an analysis of how the resource base that underpins it has likewise changed.

Another important consideration linked to production is labour. One can view labour as a resource, in the sense that without an adequate provision for it, production will be limited. The importance of labour within an African agricultural context has long been recognised, and labour-based SIs are logical considerations within sustainability. In Eroke too, labour is an important and complex factor and in the second part of this chapter labour will be examined in detail. As part of this, a discussion on the role of women in Eroke's agriculture is included as this has undergone especial change over the past few decades.

This chapter also looks at consumption patterns. Unlike other issues, consumption has received relatively little attention in studies of either agricultural sustainability or change in African agricultural systems. One notable exception is the work of Manyong and Degand (1997), and there are a number of comparative studies in a spatial rather than temporal sense (see for example, Flinn and Zuckerman, 1981). Some SI lists include vague pointers to 'nutritional status' of households and the village, but without the same detail as that provided for production and resources. This may be due in part to the western origins of the concept of agricultural sustainability, where agriculture is typically a question of profit not subsistence. Production for consumption is simply not a major issue for farmers in Europe and North America who buy food from supermarkets. Because consumption is so important to the people of Eroke it deserves inclusion.

Crops and cropping systems in Eroke

As described in Chapter 3, field and tree crops found in Eroke are typically those of the middle-belt of Nigeria. The preponderance of cassava in the cropping system is now a common sight throughout Igalaland and Nigeria. It also demarcates plot boundaries; a reminder of days when it enjoyed less prestige and an Igala farmer would have been loathe to admit cassava formed such an important part of their cropping systems and diets. Cassava is a root crop that easily slots into many cropping systems, and survives the dry season provided establishment is satisfactory. It is a long duration crop that can be intercropped in its first year (e.g. with maize, benniseed or cowpea) and then harvested the following year or even later. Also useful is its ability to produce in poor soil, often evident in a farming tradition that relegated cassava to the end of a cycle when land was due for fallow.

The prevalence of legumes is another feature in Eroke agriculture. Cowpea, groundnut and particularly bambara nut are all widely grown in the area, even though cowpea in particular suffers from pest attack. Cereals, especially maize but also millet and guinea corn, are also produced.

Symptoms of nutrient deficiency, notably nitrogen and phosphorus, are common. Other food crops such as cocoyam and yam grow close to compounds and on heavier soils near the rivers or on newly cleared land. Table 5.1 lists some yields for both sole and intercropped plots in Eroke. These yields are based on plot experiments, with intercropping combinations factored in as a treatment rather than eliminated by what can be marked over-simplification (see for example Guyer, 1997). Although these yields are relatively poor, it is possible to find worse on other soils in Igalaland.

Multiple cropping is also common. The bimodal system allows the production of two crops on the same land in a single year, with intercropping rather than sole cropping as the norm. Based on a detailed survey of 299 individual plots in 1995, the commonest mixtures are those listed in Table 5.2. Cassava and cowpea are greatest in terms of frequency of occurrence (over 60 per cent of plots had at least one of these), and the number of intercropped plots is greater than sole crop plots. Besides being an intensification of resource use, intercropping has added advantages for crop protection, compensation and competition. Eroke farmers have long since observed that intercropped cowpea has less pest attack than sole cropped although it is more difficult to spray. Many years of on-farm research in Igalaland have shown that intercropped NCVs do not generally have less yield than local varieties. Table 5.3 is an example of a Land Equivalent Ratio (LER) calculation for the maize/cassava intercrop based on small plot yields in Eroke. Even with the worst possible scenario (using the low end of the 95 per cent confidence interval) the LER for the intercrop was nearly 1.0, while with the mean yields the LERs were 2.73 (all local varieties) and 1.84 (all NCVs). Moreover farmers are fully aware of the importance of plant density and take care when sowing (McNamara and Morse, 1996). Other crops such as bennised and bambara nut, are rarely sole cropped in Eroke. Of the 217 plots that had either of these, only four were sole cropped. A typical farmer will cultivate many of the crops listed in Table 5.2, as well as others not listed (sweet potato, millet, cocoyam, vegetables etc).

The diversity of field crops is equally matched by variation in tree crops. The later are found in compounds and to a lesser extent on farms. The ethnobotany of Eroke shows sixty five tree species, some of which are still abundant while many are endangered. Oil palm is everywhere, and as seen in Chapter 3 the economy of Eroke revolves around it. Citrus abounds as does cashew, locust bean, kola nut, guava, banana/plantain and many others. Notably absent is coconut, a tree that once did well in the area. Some have plantations of coffee and cocoa, but one would expect more considering that many Eroke farmers worked on these plantations in Yorubaland.

Eroke farms give an overall impression of a typical Nigerian mid-belt cropping system, with a healthy diversity of reasonably well-maintained field and tree crops. Intercropping is often associated with a 'sustainable' cropping system, because it implies an absence of artificial inputs like fertilizer and pesticide. If that theory held, Eroke epitomises sustainability, especially if one also adds to it a biodiversity unequalled in western agri-business. But a closer look shows signs of stress, at least in terms of field crop production where yields are low. Farmers complain about depleted soils and even leguminous trees such as *Leucaena* sp. and *Gliricidia* sp., have difficulty in establishing unless some phosphate or even nitrogen fertilizer is used. The bulk of the plant nutrients are 'locked-up' in the vegetation rather than in the heavily weathered and leached soils.

Changes in crop and tree production

Change in production is often considered a key indicator of agricultural sustainability. In the course of two surveys in 1997 and 1998 farmers were asked whether aspects of crop production had changed (increased/better, decreased/worse) or stayed the same since the introduction of SAP in 1987. The following four components, of production were addressed:

1) area grown to the crop or number of stands
2) yield of the crop (produce/area or produce/tree)
3) total farm production (yield multiplied by area)
4) quality of the produce

Similar types of crop were grouped together for analysis:

(a) legumes (groundnut, cowpea, bambara nut and yam bean)
(b) cereals (maize, guinea corn and millet)
(c) 'soup' crops, typically of high monetary value (benniseed and egusi melon)
(d) root crops (yam and cassava).

The counts in each category (component of production and crop group) were analysed using Chi-square tests and a summary of the results are presented in Table 5.4 (details in Appendix A). The Chi-square tested the pattern of response between the crops in each group.

In essence, these surveys were meant to ascertain perceptions rather than come to terms with the real life situation, although it may reasonably be assumed that the two are related. Table 5.4 summarises the trend in response, but note it masks diversity of opinion. Even with a clear increase, there may be a minority who feel the trend has been in the opposite direction. Similarly, '0' which suggests no change, may in fact represent a situation where almost equal numbers answered 'increase' and 'decrease'. In a sense the '0' is nothing more than a village wide compromise rather than an unequivocal statement of 'no change'. This diversity of opinion is a key consideration, and not intended to be hidden. Although the details in the appendix are somewhat dense, it is recommended the reader take time to browse them.

The perceived pattern for production clearly varied between the different crops. In both the legume and cereal categories, the trend over the ten years was broadly a decline. Yam bean showed an increase, a trend significantly different from the other legumes. The root crops presented a different picture with cassava in particular showing a huge increase and yam less so. The two 'soup' crops, benniseed and egusi melon, were different with benniseed showing an increase in production and egusi melon showing a decline. The overall pattern clearly indicates a differentiation between the crops rather than a blanket decline in production.

The components of production (area and yield) showed similar variation. The area planted to yam bean had increased, and as farmers claimed that the crop yield had also increased this explains the perceived increase in production. This contrasts with the other legumes that show an increase in area and a decline in yield. Clearly the increase in legume area is a coping strategy to compensate for decreased yields. However, a detailed analysis of the cropping history of five farms did not show any clear increase in cowpea or bambara nut area over the ten years (Table 5.5). Despite the fluctuation in area cropped to cowpea, there was no significant difference in the proportion of farm planted to it over the years. The proportion of farm planted to bambara nut showed significant variation between years, but did not amount to a clear trend.

The trend was different for cereals, as opinion was divided over the area planted to maize, guinea corn and millet with almost equal numbers suggesting an increase and decline (hence the compromise in Table 5.4). The detailed analysis of five farms (Table 5.5) over ten years did suggest a significant decline in the proportion of cultivated land planted to maize. There was also a clear majority opinion that yields of all cereals had declined over the ten years. Those who recorded an increase in yield were specifically referring to their recent acquisition and use of fertilizer. The overall picture is one where crop area remains more or less the same (perhaps even declining) with a strong trend towards decreased yields. This accounts for the perceived decline in production of cereals. Cereal crops are often taken as indicators of soil quality, mainly because they are 'hungry' in their demand for primary nutrients especially nitrogen, phosphorus and potassium. Those aware of a decline in production saw this as signifying 'dead' or 'tired' land. However, it is not clear why more did not try to compensate for yield decline by increasing crop area, especially in the case of maize which is the more valued of cereals not only in Igala but throughout Nigeria.

The two 'soup' crops seemed to have quite different trends in production components over time. The majority in the sample suggested that areas planted to benniseed had increased with a stable yield (numbers saying 'increase' matched those saying 'decrease'). Taken as a whole, this accounts for the increase in benniseed production. Egusi melon was different, with no clear consensus as to change in land area (increase matched decrease), but more agreed that egusi melon yields had declined. These two factors explain the decline in production. Therefore the trend for benniseed indicates an increase in area with no consensus over yield, and egusi melon is the opposite. Melon is similar to cereals in that it is a 'hungry' crop with farmers interpreting decline in yield as an indicator of poor soil. However they are not necessarily willing to plant this crop on the more fertile land. The higher price received for benniseed is the primary factor in its increased cultivation.

The two root crops had a strong and similar trend towards an increase in area cultivated (clear majority of opinion in both cases). This was confirmed for cassava in the detailed analysis of five farms shown in Table 5.5. But there were differences in perception of crop yield. With cassava there was an indication of an increase in yield, primarily because of the NCVs, while opinion was divided over yam. Those who suggested an increase in yam yield maintained it was due to new land being used.

Besides production, quality of produce is a serious consideration. Quality showed varying perceived trends over the ten years. The general picture for grain crops was maintenance of quality, but with some indications that quality may have declined. Yet again there is significant variation in perception within and between crops. For example, with yam bean the balance of opinion may have tipped towards an increase in quality. There was also some suggestion that the quality of egusi melon had declined. The trend for benniseed is less certain, with equal numbers suggesting an increase and decrease and most replying 'no change'. With regard to the two root crops, opinion on yam quality was divided with almost equal numbers in all three categories. Opinion on cassava, was more weighted towards an increase in quality over the ten years; the commonly held view being NCVs were of better quality than local varieties. The diversity of opinion for yam quality was again linked to use of new land. Those not using new land saw both yam quality and quantity declining.

A similar diversity in production and quality was also apparent with tree crops. Signs of increase in numbers of palm and locust bean stands over the ten years were clear; the pattern for citrus is less clear. The same held for yield/tree, overall production and quality of produce. Emphasis on oil palm and locust bean is easily explained given their uses, particularly oil palm. The demand for palm oil continues to grow and so does the market price. Citrus fruits are not as lucrative.

Tree crop production in Eroke has been boosted by the number of women who have access to and control over oil palm trees, locust beans, banana/plantain, mango, kola nut, cashew and guava. This cultural transformation began over twenty years ago but the real impact began to be felt about fifteen years ago. Table 5.6 summarises the results of an inheritance and self-planting survey based on a sample of forty two (half men and women). Fathers answered on behalf of their young daughters. Both inheritance and self-planting of trees increased between the period before 1970 and 1990. In part, this increase is simply related to age, especially the jump in inheritance between the first two periods (178 to 1468 trees) and self-planting between the second and third periods (89 to 510 trees), but is also a reflection of changing attitudes. Since many of the inherited and planted trees are NCVs, yields are often better than some local varieties, and so is the quality. Production costs are also lower as it is easier and safer to harvest the dwarf oil palms.

What does the foregone tell us about sustainability? The pattern in production is complex. Some crops, notably the cereals and egusi melon, are clearly recognised as producing less now than in the past; other crops (yam, cassava, benniseed and yam bean) have moved in the opposite direction. The same is true for tree crops. With the five farms monitored in detail, there was empirical evidence to support the decline in areas planted to maize while areas to cassava had increased. As these trends are often cited as evidence of a decline in soil quality the situation suggests unsustainability. However, in dynamic systems change is a constant factor, and this is witnessed especially by changes in female ownership of tree crops. The increase in cassava production has been driven by the availability of varieties that are pest/disease resistant and which taste good – even as good as yam. There is clearly 'change' but does 'change' *per se* equate to 'unsustainability' or is it just compensation and evidence of innate resilience?

Soil quality

Cultivation intensities (R-factors) appear to be relatively high in Eroke (Table 5.7). Based on a detailed analysis of six farms, the mean R-factor for the whole farm (ie. the mean of individual plot R-factors that comprise each farm) varied from 32 per cent to 56 per cent. These values are high compared with the 'suggested' maximum of 15 per cent quoted by Stocking (1994) given the soils and cropping periods in Eroke. Stone's (1997) study of the middle belt in Nigeria calculated R-values (based on proportion of farm in production) at between 23 per cent and 72 per cent. The latter figure corresponds with 'intensive' farming as practised by the Kofyer people of the Jos Plateau. It is interesting that the mean R-factor for the six farms did not have a strong relationship with farm size, although two of the smallest farms in the sample did have the highest R-factors. However, there was substantial variation in R within the six Eroke farms (represented by the CV values in Table 5.7), suggesting that some plots are more intensely cropped than others. The significant inverse relationship between the mean R-value and the CV between plots suggests that more intensive farms tend to be more uniform. In general, the moisture retentive plots near compounds tend to be more intensely cultivated than the more distant ones that are less accessible and prone to theft. The variation in Table 5.7 shows the need for care when calculating R-factors for the type of farms in Eroke.

This same care is equally necessary when looking at the time dimension regarding cropping intensity. Figure 5.1 gives the percentage land under production for all the plots surveyed in the six farms. As well as differences between farms, there is also variation within farms over time. In some cases these changes can be considerable and explanations are possible. In the case of farmer A1 the year 1995 marked the end of a long period of paid employment for him resulting in a substantial and sustained increase in the proportion of land cultivated to compensate for loss of earnings. For farmer A4 the high proportion of land cultivated between 1988 and 1993 was explained by the presence of friends and family willing to help out. On his return from working in another state as a labourer in 1994, the same level of help was not available and resulted in a decline in the proportion of land under production. Farmers A6 and B3 had similar experiences. Another, A6, was smitten by ill health that resulted in a major short fall in the land cultivated in 1994. Again, poor health and family deaths account for the irregular pattern of production experienced by A9 between 1989 and 1997 after which circumstances stabilised once more. Deteriorating health has serious repercussions for farming and ultimately for income and well being, and ranks highest as an obstacle to sustainability according to most women in Eroke.

Even though the R-factors are relatively high and farmers repeatedly complained of poor soils, they are not adopting radical solutions to poor soil fertility (Table 5.8). The most common response, not reflected in Table 5.8, was to open-up new land for cultivation, with yams and yam beans enjoying preferential access to this new ground. The logical assumption would be that as new land came into use other land would enjoy a longer fallow period, but opinion differed. Some felt the fallow period had increased, but almost equal numbers felt that it had remained the same or even declined. Based on a sample of sixty six farmers, the length of the fallow period appears to be more or less constant at approximately five years between the late 1980s and the late 1990s. The mean for the late 1990s is lower than that of the late 1980s, but this was not statistically significant due to variation in opinion (Table 5.9). How can this apparent discrepancy be explained? Those asked about fallow periods tended to consider only the cultivated areas near the compound rather than the long-term fallow recently opened and usually some distance away. Their response is limited to the highly cultivated areas only, an understandable position given the ageing factor of male adults.

Other coping strategies drawn on to deal with impoverished soil were mentioned but these varied in importance. Many mentioned using compost/manure, but only on a small scale near the compound. Next to the use of new land is the increase in the area planted to legumes and to a lesser extent leguminous trees. The former was mentioned by virtually every respondent, and corresponds with views regarding the area planted to groundnut, cowpea, bambara nut and yam bean mentioned earlier.

As expected, the use of fertilizer had declined over the ten years. Given the hike in price, and major difficulties with availability, only a few farmers used more in the late 1990s than in the previous ten years. While everyone agreed that fertilizer is expensive and labour intensive to apply, all agreed it was very effective. Fertilizer use to date has not been on a scale threatening to the environment, but should farmers be able get it at the right price far more would be used. As Goldman and Smith (1995) point out for the north of Nigeria – supply of fertilizer is usually the problem not price.

Do these trends in cropping intensity leading to a decline in soil fertility say anything about agricultural sustainability? It is clear from the coping strategies adopted to deal with soil infertility and the opinion of those interviewed that the land in Eroke is dying. However, the problem with any simple interpretation is variability. For example, there is no unanimous opinion that fallow periods are increasing or decreasing. Either may be the case depending on individual circumstances. No doubt some are farming more intensely on plots near to the compound than before, and the absence of able-bodied young men and youth in the household and inadequate funds to hire labour may well be to blame. Nonetheless, some had returned to land that had been in fallow for many years, growing high value crops such as yam. A variation in the percentage of land in cultivation was noticeable, with personal circumstance contributing mainly to this situation. Other than the use of old land left to fallow, coping strategies are not particularly strong. Planting more legumes and limited use of compost in plots near the compound are certainly indicators of soil infertility problems, but hardly represent a major shift within the cropping system. Planting more cassava and less maize are also typically seen as indicative of the above, but other reasons such as the availability of cassava NCVs resistant to pests and diseases are also important. Nevertheless, the indicators point more towards unsustainability rather than sustainability.

Crop protection

Besides soil fertility, pest and diseases impact on both production and quality of produce in Eroke. All crops have some pest/disease attack, but groundnut, cowpea, maize, yam and cassava are more prone with subsequent losses in both yield and quality. Given that four of these crops are common in Eroke (at least one of these was present in nearly all of the 299 plots assessed in 1995/96) then the importance of crop pests and diseases is apparent.

Opinion pointed clearly towards an increase in pest/disease problems on all crops except cassava, over the ten years. The decline in pest/disease problems on cassava is explained by the NCVs of this crop having resistance to these problems. In some cases, the increase in pest problems was attributed primarily to the increased cost of pesticide, particularly for maize and cowpea. Even though DDS can supply, prices have increased rapidly since SAP and farmers who once applied them can no longer afford the cost. Therefore an increase in a pest/disease problem may not necessarily stem from an environmental change or a build-up of resistance. Ironically an increase in pests/diseases might suggest a trend towards unsustainability, but the situation may be the opposite if sustainability is equated with less use of pesticide.

Farmers in Eroke are far more aware of potential problems with pesticide use than they are with fertilizer. Though not common, they know of people who have died from pesticide-related problems in Igalaland and beyond. Such news travels fast. Pesticides also cause deaths in farm animals. No one practices integrated pest management (IPM) in the sophisticated forms described by Morse and Buhler (1997). As the use of pesticide is not widespread, its potential environmental damage did not arise. Therefore even the most limited form of IPM, called pesticide management by some, has not been adopted, although the high cost of pesticide does make them think carefully about its use. Even if serious pest damage occurs there is little that can be done as curative measures are limited unless the farmer has access to the required type and quantity of pesticide. In essence, Eroke practices an integrated pest control system, relying on preventative measures such as intercropping, choice of season and resistant varieties rather than curative methods or 'management'. A level of pest damage is grudgingly accepted, although farmers would readily admit that they would like all pests to disappear if possible.

In the realm of crop protection the prospects of sustainability seem more promising - less pesticide is being used than previously (although use has never been on a large scale) and the widespread adoption of resistant varieties of cassava helps. However, compared with soil quality and a host of other factors, crop protection is not the farmers' first priority except for cowpea and possibly maize.

Labour

Since men own the land, it gives them access to and control over farm produce, including trees. Farmers hire skilled labour to harvest oil palm bunches, a highly specialised and dangerous task. The first fruits are usually ready for harvesting around March. Other tree fruits have to be harvested at the same time, although processing is not as critical. The processing of locust beans can be postponed until the pressure eases. The overall picture in Eroke was one of competition for labour between the harvesting of oil palm and the cultivation of annual crops. A major labour requirement of annual crops is the need for constant weed control given the host of noxious weeds, many of which are difficult to manage by hand. A single crop may need three to four weedings to produce an economic yield.

The problem is further compounded by the erratic and unpredictable nature of the rains. No external labour is available, and there is no labour surplus in the community. Machinery services are non-existent, and draft animals unknown. This puts all the pressure on those who can work. The school holidays do not coincide with the period of high labour demand, but does for the main annual crop harvesting. Since this is mainly the realm of the women, they benefit most.

As outlined in Chapter 3, there is a marked division of labour between the sexes. Men do the heavy work of land preparation, including clearing and ridging, while women and children are mainly concerned with the time consuming but less arduous tasks of weeding, harvesting, processing and marketing (Table 5.10). As traditionally women are not involved in land preparation they hire male labour, usually labour groups, to execute these operations on their own farms (Table 5.11). Later the women hire themselves out as labour groups to do other operations (mostly weeding and harvesting), but are paid at a lower rate than their male counterparts. The men argue that operations performed by each group are different, thus justifying the higher pay for men. From the point of view of the women, men have an advantage in that women have to pay cash for the early operations on their farms forcing them to sell their palm oil prematurely, while at this time the men harvest palm bunches for sale at the best possible price. This is the time most suited to markets and local trading. The problem was somewhat ameliorated where women actually owned some tree crops, which require less labour than annual crops, and where they had a source of ready cash through the sale of unprocessed palm fruits.

A regression of total person time against land area for both male and female 'managed' plots produced a coefficient (0.695 to 0.698) that was almost identical for the two sexes (Table 5.12a). This is to be expected, as regardless of who manages the plots labour input/area should be more or less the same. Using the labour input for known plot sizes, it was possible to produce estimates on a per hectare basis. The upper and lower estimates were calculated using the 95 per cent confidence interval for the regression coefficients, and cover all the plots surveyed. As many of the plots contained cassava, and the survey covered harvesting and processing, the labour for these plots was recorded over a longer period than just a single growing season or even a year. The predicted labour inputs per hectare in Table 5.12a are therefore not on a per annum basis. To do this, the plot labour figures and their variation were used to calculate the annual labour input for the six farms mapped in detail (Table 5.7). The ranges in Table 5.7 are low and high estimates of average annual labour/ha over a ten year period for the six farms, allowing for variation in three factors:

1) labour estimate per ha in Table 5.12a (upper and lower values)
2) variation in area cultivated over the ten years
3) crops grown in each year, particularly the fact that one of the most common crops, cassava, takes longer than a year to mature.

Labour for tree crop harvesting/maintenance, path clearing and forage provision for animals are not included in these bands. The range of 350 to 600 person hours/ha/year is broadly in line with other estimates for shifting cultivators in West Africa (Stone *et al.*, 1990; Stone, 1997). The figures are at the lower end of the 480 to 960 person hours/ha/year quoted by Cleave (1974), and the 500 to 1000 person hours/ha/year given by Stone *et al.* (1990) as typical for shifting cultivators. By these standards, cropping in Eroke is by no means intensive, and may be regarded as relatively light.

The main factor that governs labour input is plot size, while the crop and the system (sole or intercrop) are of secondary importance. In general, intercrops have a higher labour input/area than do sole crops (see the notes in Table 5.12), but overall production for intercrops tends to be higher than sole crops of a comparable area. In other words more production equates to more labour. A simple illustration of this over-yielding in the maize/cassava intercrop using the LER as a tool has already been referred to (Table 5.3). A third factor influencing labour input was the presence of yam. Plots with yam in them tend to have higher labour inputs than plots without it.

The type of farm work that men and women do is qualitatively different, and distribution between the sexes does depend to some extent on who (man or woman) owns or manages the plot. As already mentioned, and as can be seen from Table 5.10, male labour is for land preparation (clearing, making heaps and ridges), planting and weeding; female labour involves harvesting (including transportation from farm to household and post-harvest activities, threshing, winnowing and storage). However, some activities overlap. Female managed plots tend to have a greater proportion of female labour, particularly for planting and weeding, but males do land preparation. One must also consider that activities are not equal in terms of time and effort. The two most intensive and time consuming activities, land preparation and weeding, are heavily reliant on male labour for both male and female managed plots. In this, Eroke is typical of Igalaland as a whole. Donors might note that the situation defies the conventional notion of African women as the 'real' farmers with men appearing to take a secondary role.

As would be expected the cost of labour for land preparation showed a significant relationship with land area, the larger the area the higher the cost, though this is not a linear relationship (Table 5.12b). Double the plot size does not necessarily mean twice the cost. This was true for the late 1980s as well as 1998. Results also suggest that men and women have to pay more or less the same for these activities per unit area of land, although the cost of labour has increased between the late 1980s to 1998. However, it seems as if the local labour inflation rate was substantially less than the official rate for the same period (1026 per cent). Given the apparent high demand for farm labour alluded to in Eroke, this is somewhat surprising, but may be due in part to a less than official rate of inflation for farm produce in the village.

Most farm labour in Eroke takes the form of group labour rather than individual or household labour, and some of the most interesting results from the labour surveys were related to group labour. The general preference seemed to be for the *adakpo*, largely because a group member can earn cash if the group works on a non-members farm (Table 5.13). This feeling however is not universal, as some respondents did not like the *adakpo* because occasionally members cheated, often failing to be present when their input was most needed. Views regarding the *ayilo* were almost the exact opposite, as the cash earning potential is zero unless a member happens to want extra work on his/her farm. However, the sole focus and onus upon members means that they come on time when required. For the *ogwu* the results were mixed, with roughly a third of respondents regarding the *ogwu* as the system most preferred while another third believed it to be the least popular (the remaining third were neutral). The *ogwu* was seen as both reliable and flexible, but expensive.

Given the way these groups operate, and individual preferences for promptness, quality of work and low cost, the ranking achieved is not surprising. Interesting too, is that membership of the *ayilo* and *adakpo* declined since the late 1980s (Table 5.14). Almost invariably this was attributed to aging which limited ability to meet membership commitments. This is a fact throughout the village as a whole, and they were keenly aware of its unwelcome repercussion and this is not unique to Eroke. Richards (1990) has made a similar observation for Nigeria and Sierra Leone.

A detailed analysis of the use of different sources of labour on plots managed by males and females did suggest some differences (Table 5.11). For male managed plots, the major source of labour was the household comprising both male and female members, and the owners were typically involved in every activity. This contrasts markedly with the female managed plots where the household labour accounted for less than a quarter of the total labour input, and the owners were less directly involved. Guyer (1997: 143) had also noted the latter point for female farmers in Idere village, Yorubaland. For female managed plots the predominant source of labour was the ogwu, accounting for well over two thirds of the total labour input; these were largely male and involved in land preparation and weeding. For both sexes, the ayilo accounted for less than seven per cent of labour input, a further reflection of its relative unpopularity. These differences between male and female managed plots are readily explained as the male managed plots comprise the main household farm. It is here that the bulk of the food and cash crops are grown, with the household drawn upon as an important source of labour. The *ogwu*, and to a lesser extent the *adakpo*, are also utilised to help ease labour shortfalls at critical times in the season. In contrast, the female managed plots are intended to provide some food for the household (mostly ingredients for the soup), but are largely a source of income for the women. Engaging some household labour, including that of the household head, comes as no surprise, but as these plots are intended to provide cash for the women the *ogwu* assumes major importance, especially for traditional male tasks.

Interpreting the labour resource base in Eroke in terms of sustainability is somewhat complex. The lack of affordable and flexible labour is a major limiting factor in production. Emigration of the youth from the village does not augur well for either its present or future sustainability, and more especially for agriculture, but other considerations can swing the pendulum back to a more positive conclusion. Farming in Eroke is not intensive and may even be regarded as relatively light work, at least in the Nigerian context. Many have also learned to put their time into less laborious, but nonetheless lucrative sources of income. Tree crops have assumed great importance with women now owning them. As labour required for tree crops is less than for field crops, and crops such as locust bean, kola nut and oil palm have a high market value, one wonders about the importance of field crops. How important is their perceived farm labour 'problem' given the time spent on a diverse range of other activities? Taking a broader livelihood context rather than just agriculture a more sustainable picture emerges. The key point seems to be the wide range of income generating options and hence livelihood resilience, a point that will be returned to later.

Firewood and water

Eroke women and children may not consider themselves hewers of wood and drawers of water, but will never deny how seriously both impact on their lives. They spend many hours collecting firewood and until recently drawing water from the stream or pursuing tankers. Water availability, consumption and quality have improved over the ten years thanks to the construction of a new rainwater harvesting tank with a capacity of 280,000 litres (Table 5.15). Collecting rain is a sustainable practice where the annual rainfall is approximately 1500mm; the tank is solidly built of concrete and plastic sheeting to ensure there are no leaks. Provided the tank is cleaned and maintained on a regular basis there is every reason to believe it will endure for fifteen to twenty years. Its contribution to community health and prevention of water borne diseases is a factor contributing to sustainability, and less work for women and children in collecting and transporting water allows them more freedom and time to engage in activities of their choice.

The situation with quality firewood is more serious as it has declined in availability over this period. There are two main categories of trees; the wild species which are best for firewood and those planted by individuals (mainly fruit trees such as mango, cashew, locust bean, guava, even oil palm). The availability of the high quality wild species is fast declining, and already women have switched to burning fruit trees, including their prized oil palm (Table 5.15; details in Appendix A). Eroke presents a classic case of unsustainability in firewood, and the price being paid is a loss of economic fruit trees. The management of this wild resource has a parallel in the animal world with the notion of a Maximum Sustainable Yield (MSY) already referred to in Chapter 1. If deaths outpace births, population will decline. Compensating by substitution with poorer quality wood is a coping strategy, but may not in itself help unless the rate of removal of the high quality species is less than the rate at which the population can regenerate. As there are no signs of prohibition of the use of these quality firewood trees, it seems reasonable to assume their decline will continue. Further clearing of the forest is, however, prohibited by the *gagos* and may be indicative of this concern for conservation of all tree crops.

Perhaps surprising in the circumstances is the fact that little use is made of teak and gmelina. Both these are fast-growing, and can provide good quality firewood. Although there are numerous small plantations of these in Eroke they are sold to timber merchants rather than used locally. The use of the older fruit trees for firewood is clearly the preferred option.

These two important resources perhaps present the starkest contrast in terms of sustainability. Water supply has many sustainable characteristics; which unfortunately is not the same for firewood. However, provided they are willing to utilise poorer quality firewood, the switch from wild to planted species of fruit tree is sustainable; a good illustration of the 'value laden' nature of sustainability.

Consumption

An examination of the foods normally consumed in Eroke throughout the year will help with an understanding and appreciation of consumption and its relevance to sustainability.

Their diet consists of starchy staples mainly derived from cassava cooked in a variety of ways. Yam, the favourite food, is now a delicacy consumed as soon as it is produced. Rice is also eaten, but more on festive and ceremonial occasions and by those who have little assistance in the kitchen or in need of light food. Sweet potato and cocoyam are cooked either as a porridge mixed with rice or boiled and eaten on its own as a snack. It can also be pounded as fufu. Since most of the staples consist of carbohydrates, protein is consumed mainly in the form of legume grains, of which the most popular is cowpea. Soups made from legume grains also include vegetables, processed locust bean, palm oil, pepper and salt and if possible dried fish. If meat were available it would be included, but most only eat meat four or five times a year. Other nutritious soups are also taken regularly. Egusi melon and benniseed are the most popular ingredients, but there are many others. These include *ogbono* made with seeds of the bush mango (*oro ayikpele*), okro and vegetables. Vegetables can include different types of spinach, bitter leaves, cocoyam leaves, baobab, *okohio* stems and occasionally cassava leaves. The latter is regarded more as a 'famine food'.

The evening meal is the main meal of the day, and is typically a combination of the staples accompanied by any of the soups. The smaller mid-day meal can also be a mixture of the above, and can include pottage made of cowpea, bambara nut, yams, cocoa yams and pigeon pea. This is not 'for filling the stomach' but to sustain the family until the main meal of the day.

The crops for household use are produced on the many plots that comprise the family farm. In addition to this women have their own plots where the proceeds belong entirely to them although the female managed plots are sometimes used to produce ingredients for soups.

In the past, consumption has not received the attention it might within agricultural sustainability. To redress it, this study asked the following two questions:

152 Visions of Sustainability

1) is there evidence of a change in consumption patterns (over the ten years of the study)?

2) how does consumption match production?

In addressing the first question, women from twenty households were asked about change in their consumption of foodstuffs between the late 1980s and 1998 (Table 5.4). The same respondents were also asked about their household consumption in 1997. The second question was more complex for as household size and composition change so does consumption. However, given that information was available for crop yields in Eroke (Table 5.1), as well as a detailed cropping pattern for six of the households in 1997, it was possible to examine how production matched consumption for these. Focusing on just one year helped a detailed analysis, but even then simplification was inevitable.

For the most part there was no clear trend in crop consumption over the ten years, except for groundnut, bambara nut, yam and cassava. Cassava consumption increased, while consumption of other crops decreased. A decline in yam consumption allied to an increase in cassava consumption certainly matches the prevailing view that cassava was gradually replacing yam as the staple root crop in Igalaland. But what about the increase in yam production noted earlier? How can this be possible if consumption is perceived to be declining? One obvious answer is the increase in yam price that motivated farmers to produce this crop, at least in sufficient quantity for their own use. The general impression was that crop production went a long way to meeting needs.

Consumption of meat and animal products, mostly eggs, shares the same variation as all other commodities over the ten years. Bush meat, goat and sheep are the most common sources. However, there has been a clear increase in consumption of the cheapest and most available sources of protein: chicken and eggs. One reason for this may be the DDS activities in the area that put much emphasis on nutrition. The provision of adequate protein in Igala diets has been its particular concern for many years. Crops such as cowpea, groundnut, soybean, and the provision of cheap meat, including at one time, rabbit raising were promoted. In contrast, other sources are relatively expensive, and consumed only at special occasions, social events, funerals, Christmas and Sallah. The two rarest sources, guinea fowl and bush meat, have declined in consumption. These are wild sources not seen nowadays. Indeed all the indications are that they are buying meat and eggs rather than producing them locally. As can be seen in Table 5.4 the ownership of goats and hens had declined between the late 1980s and 1998, largely due to greater incidence of diseases. Why this should be so is not entirely clear, although some blame the high cost of medicine for treating the animals.

Average household consumption of various foodstuffs in a year is presented in Table 5.16. The variation between households was significant, and is partly a reflection of household composition (number of people, proportion of adults etc.) although relative prosperity is also important. As would be expected, cassava accounts for a large proportion of energy intake, but maize, rice and yams are also significant contributors. The importance of cowpea as a source of protein is also noted.

An attempt to match production with consumption is illustrated in Table 5.17 (detailed calculations and explanation of method in Appendix B). When compared with the known area of each crop cultivated in 1997, the match is good with seventeen out of the twenty-six calculations at least equaling the actual area cultivated. However, there were a number of examples where the area cultivated fell below that needed to sustain the level of consumption. This was the case for cowpea in four out of the five households concerned. The likelihood is that cowpea production by these households is not adequate to sustain consumption suggesting the shortfall had to be purchased. This is what the women said at the commencement of this study, and adequate production of cowpea is clearly a problem in Eroke. A lively trade in cowpea is noticeable in Eroke, a crop often regarded as one of the most profitable of field crops. Although common in the area, it is severely limited by pest attack, perhaps more than any other crop in Igalaland. As a result, yields are typically low and variable. The remaining five instances where the area fell below that predicted for consumption were for benniseed, maize (two occasions each) and cassava (once).

What can these consumption patterns tell us about agricultural sustainability in Eroke? There had been little qualitative change over the ten years, apart from an increase in consumption of cassava and a reduction in yam. Some tried to improve the protein content of their diets as seen in their perceived increase in consumption of chicken and eggs. Far more apparent is the widespread cultivation of soybean. In 1997 production seemed to broadly match consumption, but the picture is complicated because households readily buy food to supplement what they produce. There is no 'non-food' crop such as cocoa, coffee or cotton, so much of what they grow they can consume; a picture that contrasts with some Yoruba households (Flinn and Zuckerman, 1981). Children are well fed, and malnutrition is not a general problem. Nobody mentioned hunger, and the consumption of classic hunger foods (e.g. cassava leaves) existed only on a minor scale. The consumption picture is one of sustainability.

Are production and consumption sustainable?

The impression created in Chapters 3 and 4, is that to all appearances most Eroke people believed in team work with men and women co-operating in many tasks. Household members operate as a team with the husband and wife producing a quick maturing crop for ready cash. This is often to offset school fees, a responsibility of the husband. However to meet the increase cost in education both work hand in hand to solve their family problems. This amounts to a third tier in the household structure, for previously the men usually provided the cash requirements in the households, but when shortfalls occur, the women are expected to assist. Both recognise the situation is now different and calling for this new development. Some other re-alignments of traditional demarcation lines have occurred, with men engaging in what was formerly a women's role, i.e. trading or producing of vegetables, and women owning tree crops. The above examples highlight the constant change necessary for survival and sustainability.

Determining what production and consumption patterns contribute to sustainability in its 'western' sense, is far less clear. A summary of conclusions drawn from this chapter is set out in Table 5.18, with the problem clearly highlighted. In four of the areas listed in the table there is evidence of sustainability, yet in three of the important areas (notably production, soil quality and firewood) there is unsustainability. If one takes a narrow view and focuses solely on crop production and the soil resource that underpins it, one could argue that the village is in trouble. A decline in soil fertility and concomitant changes in crop production gave rise to coping strategies, but to what extent can these compensate and will there be a severe collapse in production? It is difficult to provide an answer for a future based on information collected to date, but what is clear is that over the past ten years there has been change without collapse. Firewood trends give some cause for concern. All indicators point towards unsustainability, yet people are coping by using less preferred trees, including those as valuable as oil palm and other economic trees planted by themselves.

In complete contrast, water consumption, availability and quality have increased as a result of DDS's catalysed rain harvesting project, and given its inherent sustainability there is room for optimism. There are more positive trends with consumption, and perhaps to a lesser extent with crop protection and labour. There are indications that production is able to meet at least most consumption needs, and shortfalls (especially with cowpea) can be readily imported. Labour shortage for farm work is easy to understand, given the many important off-farm activities from which they derive income. Yet given the income from these off-farm activities there ought to be sufficient funds to employ labour - if it were available.

How does one balance unsustainability in some areas against obvious sustainability in others? How is unsustainability in quality firewood balanced against sustainability in water supply? Is this balancing necessary? While Table 5.18 may suggest unsustainability, it is vital to remember we are not dealing with a static system, and change infuses all boxes in the table. Individual circumstance is important as a driving force for change, and is sometimes easily missed. So far it has been evident that households have not been constrained in terms of their options, as people readily get involved in activities not directly related to crop production. Seeing sustainability in terms of the areas highlighted in Table 5.18 is immediately misleading as each of the boxes has to be seen within a wider context of livelihood. It is this broader context that will be examined next.

Table 5.1 Mean yields and standard deviations (SD) for some common crops found in Eroke

Crop	Var.	System	No. of plots	Mean yield (kg/ha)	SD (kg/ha)
cowpea	L/N	S	16	278	82
rice	L/N	S	12	613	358
benniseed	L	I	6	68	45
bambara nut	L	I	6	378	181
maize	L	S	7	571	
	L	I (cassava)	14	700	391
	N	S	7	1014	
	N	I (cassava)	14	825	
cassava	L	S	7	5714	
	L	I (maize)	14	8599	5638
	N	S	7	9087	
	N	I (maize)	14	9382	
maize	L	S and I	21	636	391
	N	S and I	21	920	
cassava	L	S and I	21	7157	5638
	N	S and I	21	9235	

Cowpea and rice data based on sole crop trials conducted in 1987 (cowpea) and 1994 (rice). Each trial consisted of four treatments: local variety (L) and three new crop varieties (N). The SD's were calculated for the pooled data (i.e. site and treatment effects ignored). Maize/cassava data are based on seven trials (randomised block design with eight treatments) conducted between 1992 and 1996. Treatments were:

(a) sole plots (S) of local (L) and new crop varieties (N)
(b) all four combinations of the crops/varieties grown as intercrops (I).

Maize/cassava yields were analysed with ANOVA. Means and SD's are expressed in terms of system yields as well as variety yields pooled over systems. All plots were farmer-managed.

Table 5.2 Some crop mixtures in Eroke

Crops	cass	c'p	b'n	b's	y'b	maize	melon	yam	g'c
cass	**57**	57	16	39	4	11	1	0	1
c'p		**22**	15	60	1	31	0	0	1
b'n			**4**	35	0	0	0	1	0
b's				**0**	0	11	1	0	0
y'b					**1**	1	1	13	5
maize						**1**	0	1	0
melon							**2**	10	0
yam								**6**	3
g'c									**1**

The total number of plots assessed was 299 (1995/96 labour survey). Figures represent the number of plots in which that pair of crops were found together (some intercrops could have up to four crops). A crop matched against itself (bold type) represents the number of sole crop plots.

Crop	no. of plots	% of total plots
c'p	187	63
cass	186	62
b's	146	49
b'n	71	24
maize	56	19
yam	34	11
y'b	26	9
melon	15	5
g'c	11	4

Total number of sole crop plots = 115 (94 of them listed in the above table)
Total number of intercropped plots = 184

cass = cassava; c'p = cowpea; b'n = bambara nut; b's = benniseed (sesame)
y'b = yam bean; g'c = guinea corn (sorghum)

Table 5.3 Calculation of the partial Land Equivalent Ratio (pLER) and Land Equivalemt Ratio (LER) for the maize/cassava intercrop in Eroke

Crop	Var.	Sole crop Mean	Sole crop CI	Intercrop Mean	Intercrop CI	pLER based on means	pLER 'worst case'
maize	L	571	301	700	213	1.23	0.56
	N	1014		825		0.81	0.47
cassava	L	5714	4337	8599	3066	1.5	0.55
	N	9087		9381		1.03	0.47
intercrop	all L					2.73	1.11
LER	all N					1.84	0.94

Data are from 'farmer managed' plots planted in Eroke (Table 5.1). Yields are in kg/ha.

CI = 95% confidence interval
L = local variety
N = new crop variety

Table 5.3 continued

LER Calculations

$$pLER = \frac{\text{intercrop yield}}{\text{sole crop yield}}$$

$$LER = \frac{\text{maize intercrop yield}}{\text{maize sole crop yield}} + \frac{\text{cassava intercrop yield}}{\text{cassava sole crop yield}}$$

A LER greater than 1.0, or a pLER value greater than 0.5 suggests, 'over-yielding' in the intercrop.

The LER and pLER have been calculated in two ways.

1) using the mean yields for sole and intercrops.

2) 'worst case' calculation by taking the mean intercrop yield minus the CI and the mean sole crop yield plus the CI.

Table 5.4 Perceptions of agricultural change over a 10 year period (late 1980s to 1998)

Crop	Area/no. of stands or animals	Yield	Prod.	Qual.	Pests/ diseases	Con.
groundnut	+	-	0	-/0	++	-/0
cowpea	+	-	-	-/0	++	0
bambara nut	+	-	0	-/0		-/0
yam bean	++	+	+	0		0
Sig.	*	***	**	*		ns
maize	0	--	-	-/0	+	0
guinea corn/ millet	0	--	-	-/0		0
Sig.	*	ns	ns	ns		ns
benniseed	++	0	+	0		0
melon	0	-	-	-/0		0
Sig.	**	**	***	**		ns
yam	++	0	+	0	+	-/0
cassava	++	+	++	+/0	--	+/0
Sig.	ns	***	***	**		***
citrus	0	0	0	0		
oil palm	++	++	++	+		
locust bean	++	+	+	0		
Sig.	***	**	***	***		
goats	-				++	
hens	--				++	
Sig.	ns				ns	
'wild' meat						-/0
goat/sheep						0
chicken/egg						+/0
Sig.						***

Table 5.4 continued

Most of the table is based on the results of the 1997/98 production surveys (total of 40 male and 33 female respondents). Consumption trends were based on a survey conducted in 1998 (20 female respondents).

prod.	production (yield by area or stands)
qual.	quality
con.	consumption
'wild' meat	guinea fowl and bush meat
0	no discernible trend over period
- and +	moderate decrease and increase respectively
-- and ++	strong decrease and increase respectively
-/0 and +/0	no clear trend, but perhaps a tendency towards decrease and increase respectively

The above represent a 'balance of opinion' amongst the respondents, and there may be marked differences of opinion between them.

ns = not significant at the 5% probability level; * $P < 0.05$; ** $P < 0.01$; *** $P < 0.001$

Table 5.5 Crop areas for five farms over the period 1988 to 1997

The data are the mean areas of the crop (over the five farms) expressed as a percentage of the total farm area in production. Also shown are the standard deviations (SD) of the percentage areas for the five farms each year, and the results of an ANOVA test applied to the percentage crop area data.

Year	cowpea Mean	SD	bambara nut Mean	SD	benniseed Mean	SD
88	22.7	9.8	14.7	10.1	16.3	9.1
89	17.6	10.8	3.4	3.7	14.1	16.1
90	22.9	20.4	5.2	7.6	9.7	7.6
91	25.1	18.9	3.6	7.7	8.8	7.1
92	21.8	18.7	0.6	1.3	5.8	9.5
93	22.4	20.0	9.0	8.5	10.9	12.9
94	19.0	12.9	4.0	2.6	9.8	7.4
95	20.1	13.6	1.6	3.5	3.6	5.4
96	14.6	9.2	8.8	8.2	4.5	8.4
97	15.9	11.1	5.6	4.3	9.8	11.1
Mean	**20.2**		**5.6**		**9.3**	
SE	**2.16**		**0.86**		**1.41**	
CV	**76**		**108**		**107**	
Farms	ns		ns		ns	
Years	ns		*		ns	
Trend	ns		ns		ns	

Table 5.5 continued

Year	maize Mean	SD	cassava Mean	SD
88	30.1	12.2	34.3	15.2
89	29.6	18.4	48.6	28.0
90	34.9	15.5	58.8	25.1
91	27.9	23.6	47.3	26.9
92	36.9	14.1	35.8	21.9
93	32.4	29.8	52.3	34.5
94	29.0	24.2	47.4	35.0
95	20.3	15.8	69.1	16.4
96	22.5	20.5	58.8	26.9
97	13.7	13.6	74.0	26.2
Mean	27.7		52.6	
SE	1.81		2.71	
CV	46		36	
Farms	***		***	
Years	ns		*	
Trend	* (-)		** (+)	

farms df = 4
years df = 9
error df = 36

The standard error and coefficient of variation (CV) are based on the error ms of the ANOVA. Trend refers to the linear trend over years based on regression analysis (- = decrease over time; + = increase over time).

ns = not significant at the 5% probability level; * P < 0.05; ** P < 0.01
*** P < 0.001

164 Visions of Sustainability

Table 5.6 Tree crop inheritance patterns for women in Eroke (before 1970 to 1990)

Period	Tree	Inherited from: father	Inherited from: husband	**Total**	Self-planted
bf 70	citrus	6	6	**12**	18
	locust bean	124	0	**124**	16
	oil palm	40	0	**40**	0
	mango	0	0	**0**	0
	kola nut	0	0	**0**	0
	cashew	0	0	**0**	0
	guava	0	0	**0**	0
	banana/plantain	2	0	**2**	0
	Totals	**172**	**6**	**178**	**34**
70-80	citrus	44	4	**48**	16
	locust bean	46	107	**153**	24
	oil palm	580	600	**1180**	0
	mango	32	9	**41**	6
	kola nut	5	11	**16**	3
	cashew	0	4	**4**	0
	guava	0	0	**0**	0
	banana/plantain	17	9	**26**	40
	Totals	**724**	**744**	**1468**	**89**
80-90	citrus	91	47	**138**	69
	locust bean	159	180	**339**	55
	oil palm	148	259	**407**	69
	mango	69	62	**131**	41
	kola nut	13	11	**24**	16
	cashew	105	82	**187**	84
	guava	1	13	**14**	10
	banana/plantain	99	67	**166**	166
	Totals	**685**	**721**	**1406**	**510**

Number of respondents = 42 (21 men and 21 women).

Table 5.7 Average cultivation intensities (R-factors), standard deviation (SD) and coefficient of variation (CV) for six farms in Eroke

Farm (code)	No. of plots	Farm area (m^2)	Mean R-factor (%)	SD (%)	CV (%)	Labour input (hours/ha/annum)
A3	36	46,125	56	12	21	373 – 535
A4	19	43,402	52	15	29	359 – 520
A6	36	81,645	40	12	30	387 – 552
B3	44	125,641	38	13	34	405 – 572
A9	39	34,647	36	24	67	435 – 606
A1	52	87,177	32	23	72	382 – 545

R-factors have been calculated on the basis of seasons not years. There are two seasons/year and a total of 10.5 years (1988 to early season 1998), hence the total number of growing seasons is 21.

$$\text{Cultivation intensity (R-factor)} = \frac{\text{Seasons under cultivation}}{\text{Total number of seasons (21)}} \times 100$$

R-factors were first calculated for each individual plot of the farm over 21 seasons, and these were then used to produce an average R-factor for the whole farm. The SD represents the variation between the R-factors for the plots of the farm used to produce the average (i.e. the SD represents within farm variation). The CV is the SD divided by the mean. Note that the CV is inversely related to the mean R-factor (linear regression after logarithmic transformation of both variables is significant at 5%).

The labour values are predicted ranges (lower and upper values) for each farm over a ten year period based on recorded labour inputs/plot in Table 28 and the cultivation patterns of the six farms.

Table 5.8 Perceived change in adoption of coping strategies for dealing with problems of soil fertility between the late 1980s and 1998

Strategy	Change
fallow periods	0
fertilizer use	--
compost/manure	++
burying of crop residue	0
planting of leguminous trees	++
planting of leguminous crops	++
Significance	***

Notes as for Table 5.4 (details in Appendix A). Results collected during the production surveys of 1997 and 1998 (40 male and 33 female respondents).
*** $P < 0.001$

Table 5.9 Change in length of fallow period

Time	Mean (years)	SD (years)	N
late 1980s	5.13	3.22	66
late 1990s	4.63	2.38	66
	ns (error df = 130)		

ns = not significant at the 5% probability level

Mean fallow length (years) and standard deviation (SD) for the late 1980s and 1990s. Results collected during the production surveys of 1997 and 1998 (40 male and 33 female respondents in the survey, but only 66 chose to answer this specific question).

Table 5.10 Male and female involvement in farm activities

Activity	Mean hours/plot	Male managed plots		Female managed plots	
		Male	Female	Male	Female
land preparation	83	100	0	95	5
planting	20	86	14	52	48
weeding	63	100	0	74	26
harvesting	44	38	62	14	86
post-harvest	18	13	87	0	100

Based on an analysis of 299 plots in 1995/96. Figures are the percentages of labour for that activity contributed by males and females. The mean person/hours/plot have been calculated over all 299 plots in the survey, and are intended to provide the reader with an idea of scale.

Table 5.11 Sources of farm labour for male and female farmers

Source of labour		Male managed plots	Female managed plots
household		44.0	23.1
help		0.2	2.2
hired (daily rate)		0.1	0
groups	*ayilo*	6.9	5.7
	ogwu	29.8	69.0
	adakpo	19.0	0
totals		100	100

Based on an analysis of 301 plots in 1995/96. Figures are the percentages of the total labour coming from that particular source.

Table 5.12 Two sets of analyses designed to determine the relationship between labour input with land area

(a) Labour input (person hours) related to land area (m^2).

Sex	N	Regression (sig. and error df)		Regression coefficient		Predicted labour input/ha	
		intercept calculated	intercept = 0	value (b)	SE	upper value	lower value
M	217	*** (215)	*** (216)	0.698	0.006	692	555
F	82	*** (80)	*** (81)	0.695	0.01	724	501
both	299	*** (297)	*** (298)	0.697	0.005	673	560

Data from the 1995/96 labour survey (N = number of respondents), and refer to plots planted to a range of crops and systems (sole and intercrop). Male and female refer to plots owned/managed by males and females.

(b) Labour costs (Naira) related to land area (m^2) for the late 1980s and 1998.

Sex and period	Regression (sig. and error df)		Regression coefficient		Predicted labour cost (Naira/ha)	
	intercept calculated	intercept = 0	value (b)	SE	upper value	lower value
(a) Male						
Late 80s	ns (11)	*** (12)	0.792	0.029	2630	824
1998	* (12)	*** (13)	0.928	0.022	8017	3311
(b) Female						
Late 80s	* (11)	*** (12)	0.838	0.03	4093	1236
1998	* (13)	*** (14)	0.932	0.035	10666	2679
(c) both						
Late 80s	** (24)	*** (25)	0.814	0.021	2679	1213
1998	* (27)	*** (28)	0.93	0.02	7656	3597

Table 5.12 continued

Notes for Table 5.12

In Table 5.12b labour only refers to land preparation (clearing and ridging) and weeding. Labour inflation rates (%) for the 10 year period (late 1980s to 1998) were 873, 763 and 531 for male, female and both respectively (maximum figures based on the predicted costs/ha). Official consumer price inflation between 1988 and 1995 was 1026%. In both analyses the logarithm (base e) of area (in square metres) was the independent variable and logarithm (base e) of the labour input (person hours; first analysis) and the cost of labour (Naira; second analysis) was the dependent variable. The significance of the regression (* $P<0.05$); ** $P< 0.01$; *** $P< 0.001$) and error df (in parentheses) have been provided. The regression coefficients (and standard errors of the coefficients; SE) were calculated based on the assumption that the intercept was zero. Upper and lower estimates of labour cost/ha and labour input/ha were obtained using the 95% confidence limit for the regression coefficients.

Based on a General Linear Model (GLM) analysis of covariance of labour, with plot area as the covariate (labour and areas transformed by taking logarithms), the main determinants of labour input were:

(a) plot area ($P<0.001$; df = 1)
(b) intercrop or sole crop (P=0.061; df = 1)
(c) the presence of yam (P=0.064; df =1)

Error df for the GLM was 295 (number of plots = 299). The adjusted means (transformed and based on a mean log area = 7.32) and standard errors (based on the adjusted mean square of 0.395) are as follows:

Sole crop plots = 5.132 (0.059)
Intercrop plots = 5.274 (0.046)
Plots with yam = 5.323 (0.121)
Plots without yam = 5.084 (0.038)

Table 5.13 Preferences in 1998 when using labour for farm work

Type of labour	Least preferred	Most preferred	Sig. (df)
ayilo	16 (5.5)	5 (-5.5)	** (2)
ogwu	11 (0)	11 (0)	
adakpo	5 (-5.5)	16 (5.5)	

Total of 32 respondents, although not all expressed an opinion regarding labour group preference. ** $P < 0.01$

Table 5.14 Membership of labour groups in the late 1980s compared with 1998

| Group | Period | Membership of group | | Sig. (df) |
		no	yes	
ayilo	late 80s	9 (-6.25)	22 (6.25)	** (1)
	1998	22 (6.25)	10 (-6.25)	
adakpo	late 80s	5 (-6.81)	26 (6.81)	*** (1)
	1998	19 (6.81)	13 (-6.81)	

Total of 32 respondents. ** $P < 0.01$; *** $P < 0.001$

Table 5.15 Perceived change in firewood and water consumption and availability (1988 to 1998)

Resource	Quality	Availability	Distance of collection	Consumption
firewood (all)		--		++
(a) planted		++	0	
(b) not planted		--	-	
Sig. (ab)		***	***	
4 high quality species		-		
4 low quality species		0		
Sig.		***		
water	++	++		++

Notes as for Table 5.4 (details in Appendix A). *** $P < 0.001$

Table 5.16 Household (HH) size and HH consumption of foodstuff (kg/annum)

Item	No. of respondents	Sample mean (kg/annum)	Sample SD	Sample CV (%)
household size	20	15	7	47
rice	19	229	136	59
cowpea	20	650	630	97
bambara nut	12	97	126	130
benniseed	20	55	31	56
melon	19	33	35	106
maize grain	20	669	417	62
cassava tubers	20	4318	2002	46
yam tubers	20	939	449	48

Mean consumption, standard deviation (SD) and coefficient of variation (CV, %). Twenty families were questioned, but not all families mentioned a particular foodstuff. Cassava and yam tuber weights refer to fresh tubers.

Table 5.17 Predicted land requirement and actual areas cultivated in 1997 for a number of the staple crops in Eroke

Household size	Crop	Predicted land requirement (ha) Lower	Upper	Area cultivated in 1997 (ha)
14	rice	0.182	0.644	**1.359**
	cowpea	1.414	3.528	**2.444**
	bambara nut	0	0.77	**0**
	benniseed	0.252	3.374	0
	maize (L)	0.532	1.568	**0.922**
	maize (N)	0.392	0.98	
	cassava (L)	0.238	0.938	**0.324**
	cassava (N)	0.196	0.644	
13	cowpea	1.313	3.276	0.781
	bambara nut	0	0.715	**0.333**
	benniseed	0.234	3.133	**0.466**
	maize (L)	0.494	1.456	**1.113**
	maize (N)	0.364	0.91	
	cassava (L)	0.221	0.871	0.013
	cassava (N)	0.182	0.598	
22	cowpea	2.222	5.544	0.48
	bambara nut	0	1.21	**0.467**
	benniseed	0.396	5.302	**0.511**
	maize (L)	0.836	2.464	0.478
	maize (N)	0.616	1.54	
	cassava (L)	0.374	1.474	**1.062**
	cassava (N)	0.308	1.012	
7	cowpea	0.707	1.764	0.543
	bambara nut	0	0.385	**0.19**
	benniseed	0.126	1.687	0
	maize (L)	0.266	0.784	**0.635**
	maize (N)	0.196	0.49	
	cassava (L)	0.119	0.469	**0.744**
	cassava (N)	0.098	0.322	

Table 5.17 continued

Household size	Crop	Predicted land requirement (ha) Lower	Upper	Area cultivated in 1997 (ha)
13	cowpea	1.313	3.276	0.438
	bambara nut	0	0.715	**0.374**
	benniseed	0.234	3.133	**1.297**
	maize (L)	0.494	1.456	0.173
	maize (N)	0.364	0.91	
	cassava (L)	0.221	0.871	**0.569**
	cassava (N)	0.182	0.598	

Bold type refers to when the actual area cultivated is within or above the predicted range. Rice is a marginal crop on much of the Eroke soils, and only the first farmer in the list had suitable soils on which to grow it. Estimates are based on local varieties (L) except where new crop variety (N) is indicated.

Table 5.18 A summary of production and consumption based indicators in terms of sustainability

Area of concern	Summary	Some indicators
production	unsustainable (but marginal)	decline in cereal production legume production constant increase in cassava, benniseed, yam and some tree crop production
soil quality	unsustainable	old fallow used for yam production more area planted to legumes and cassava (but not sure if linked solely to soil fertility) less area planted to maize and lower yields
crop protection	sustainable	less pesticide used more pests on some crops but linked to less pesticide being used less pests on some crops (eg. cassava)
labour	sustainable (but complex)	some evidence of labour shortage for farm work labour employed in a wide range of off-farm activities
consumption	sustainable	production seems to at least match consumption for many crops (notable exception of cowpea) no evidence of malnutrition or hunger
firewood	unsustainable	increase in consumption decrease in availability of high quality firewood switch to less preferred, including planted, trees
water	sustainable	increase in consumption and availability increase in quality

(a) farm A3

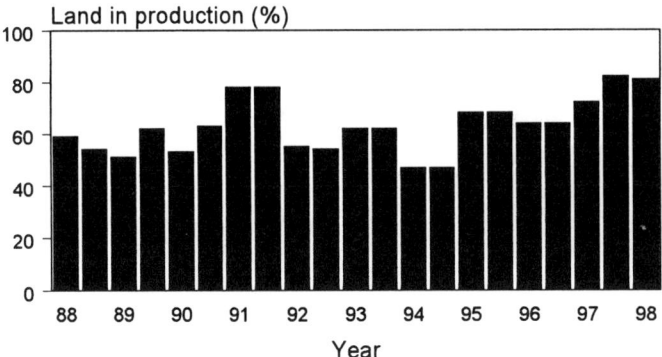

(b) farm A4

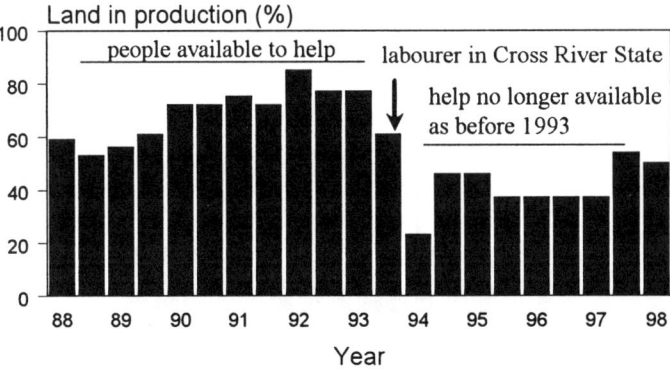

Figure 5.1 Proportion of land under cultivation for six Eroke farms

(c) farm A6

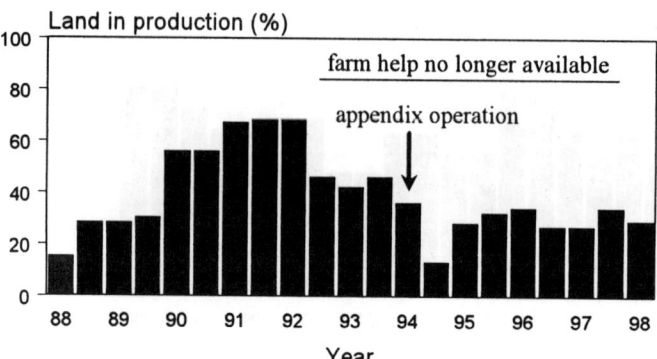

(d) farm B3

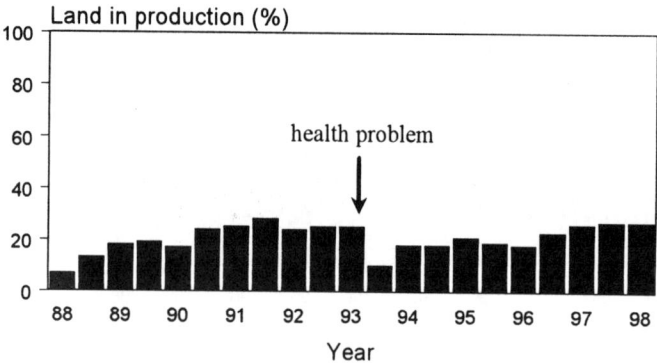

(e) farm A9

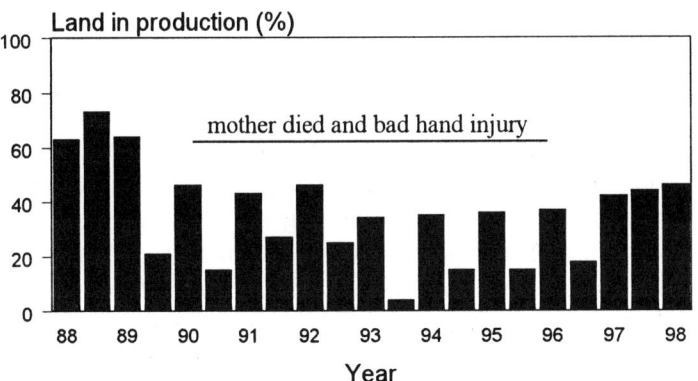

(f) farm A1

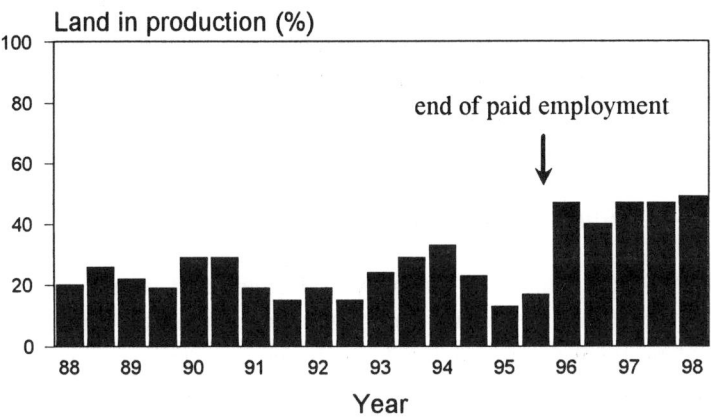

6 Livelihood and Leisure

Introduction

As seen in the previous chapter, agriculture in its wider livelihood sense (includes crop/animal production and tree crops) is undoubtedly an important axis around which the lives of Eroke inhabitants revolve. Their evenings are not spent discussing agriculture only for Nigeria furnishes them with an ample supply of topics. They speculate on decisions that encroach on their lives, but music, song, dance, story telling and partaking of local drink help pleasantly while away many an evening. Leisure is an important side to their lives and one has to appreciate that to fully understand livelihood and sustainability (Guyer, 1992; Lawuyi and Falola, 1992).

This chapter looks at the broader picture of livelihood and leisure, and as part of this, the formal and non-formal institutions in Eroke will be explored. Despite its dependence on indigenous institutions, the people have an appreciation of western styles, especially education and medicine. Both are seen as complimentary, but unfortunately what might be an ideal merger is undermined by the vagaries of SAP and the pursuit of white and blue collared jobs in the urban areas. Finance plays a central role in their livelihoods, and this includes the socio and cultural traditions that bind households and their members into a dynamic village community.

Occupation and sources of family income

Males in Eroke see themselves primarily as farmers and women as traders. However, both felt that their other occupations were important, and this was particularly true for women. Table 6.1 illustrates the number of occupations and sources of income claimed by male and female respondents. None of the differences between the sexes are statistically significant, but the majority of people claimed to have at least three occupations and four main sources of income. Women seemed to have more sources of income than men, although the average age of women interviewed was significantly lower than that of the men so it is possible that young people are simply more adventurous.

As could be expected, men ranked agriculture higher as a source of income than women (Figure 6.1). Cowpea, benniseed and egusi melon enjoy pride of place both as cash crops and in household consumption. Tree crops, particularly the oil palm and its abundance of by-products, were also highly rated as an income source by both males and females.

Trading plays a major role in Eroke, with both males and females ranking it highly as a source of income (Figure 6.1). This mostly involves the purchase of foodstuffs from neighboring producers which are then transported to other markets for resale. Some markets may be local, for example Imane, Ankpa, Okpo and Ogugu, but more are in the North of Nigeria as far away as Kano. Table 6.2 presents the results of a survey targeted at known traders in Eroke. All the women interviewed regarded themselves as long-distance traders, but half of them also saw themselves as short-distance traders. Only two male traders were interviewed – largely because men did not apply this category of occupation to themselves. Even so, these two men did not see themselves as food traders although they were involved in the wholesale marketing of food ingredients as well as medicinal artifacts and animals. The message is that although produce originates in Eroke, the horizons of its traders can extend far beyond the village boundary.

Other more diverse activities provide important sources of supplementary income. These include services such as hairdressing, sewing, repair of tools, motorcycles and other machines, as well as the selling of prepared foods (all included in the 'others' category in Figure 6.1). Many of their customers come from outside the village, again illustrating the complex web of inter-relationships that link Eroke with the wider world.

Besides production and trade, people in Eroke enjoy an income from remittances. Although this is not a major source of income, remittances are socially important in maintaining a link between emigrants and home. Remittances can take the form of physical assets such as clothes and food as well as cash, and all members of the household benefit (Table 6.3). Also important are new ideas and skills which 'abroadians' acquire and practice on their return to the village. These are highly valued in Eroke, and mirrors the experience of McNamara and Morse (1998) who found that of all the services offered by DDS, 'networking' and the provision of new ideas were highly esteemed by Igalas. An example is shown in Table 6.4. Here a number of male key resource people that have had contact with a returning migrant worker were asked to list the benefits they had received. Although the survey

180 Visions of Sustainability

had relatively few respondents, a clear majority espoused the benefits of new knowledge and skills compared to the more 'physical' items such as planting material and agro-inputs.

Family expenditure

Health care (including medicine), food and school fees (Figure 6.1) respectively constitute the three major items of expenditure as agreed by males and females. Food importation is now a necessity, hence its prominence in the list of expenditures. The cost of food is covered by income from activities such as trading and provision of services to customers within and outside the village. Therefore the purchase of food is sustainable provided these activities continue; there was no indication to the contrary. Farming ranked forth and sixth by males and females respectively, and this confirms the findings of others that point to a preference for investment in off-farm activities in Nigeria (Flinn and Zuckerman, 1981).

The importance of health and education has long been known in Eroke, as well as elsewhere in Nigeria (for example see Guyer, 1997). Both these costs have increased since the late 1980s, and Table 6.5 illustrates this for the health care and education institutions near Eroke. The variation in the percentage increase for the various institutions is largely a reflection of popularity, itself a function of reputation and price. For example, the 'Winners' Primary School, a popular private school in Imane, charges high fees from which it pays salaries, while the government financed schools charge less because government pays the salaries. With the exception of a few substantial increases, the general picture suggests an increase less than the official inflation rate for the period 1988 to 1998 or even a more local estimate based on a series of locally important items (Table 6.6). Regardless, money has to be found to match the difference. Attendance at primary school is compulsory for all children in the required age group, so the costs are unavoidable. More flexibility is apparent in the three important sources of health care in Eroke (Table 6.7), of which only the 'clinic/hospital' group is costed in Table 6.5. The most popular option is the DDS run clinics with food and nutrition demonstrations. Adult women and children in particular attend clinics (hence the significant difference in Table 6.7). Herbalists have assumed a renewed popularity in Eroke, and the increase in price of western medicines has forced the population at large to reconsider traditional herbal treatments.

Indigenous institutions

Of the many important indigenous institutions in Eroke, only the *oja* will be discussed in this chapter. There are eight *ojas* in Eroke at present (Table 6.8), and their striking feature, with the exception of St. Cecilias, is that membership is mixed. In most other places in Igalaland *oja* membership divides strictly along gender lines. The objectives of all *ojas* are similar, though variations exist regarding frequency of meetings and disbursement of contributions. All *ojas* have a constitution, a live register of members and a report on all disbursements. Records, held by secretaries are available to members for examination either on meeting days or by request. However, this rarely happens as most members are illiterate. *Ojas* are usually named according to the day of the week or market day on which they occur, the exceptions being St Cecilia and *oja Udama* (means 'unity' in Igala). Six of the Eroke *ojas* have a weekly contribution arrangement, and two (*Udama* and one of the *Lami*) are based on a more complex share system.

Udama is the overall executive oja, with representatives from five of the other ojas; *Lami, Igeya, Ladi, Lami* and *Afor*. They are part of a group of twenty-five *ojas* from Imane district that holds shares in an Ankpa bank. The formation of this group was encouraged by the 'Better Life Programme', an initiative of the Babangida regime introduced when Igalaland was part of Benue state. Members of the Udama purchase 'shares' at N50 per share, and each member can have more than one. *Udama* has a total of 120 shares, and the bank in Ankpa gives a loan to the *Udama* based on their share capital (number of shares multiplied by unit share value). The loan received by any shareholder is reckoned on the basis of the number of shares held. However this group seemed to have secured two grinding mills, one for maize and another for gari (cassava) together with a jack, before they had any savings. Two workers paid by the *Udama* operate these in the main market in Imane.

Lami is also based on a 'share' system, but works differently to *Udama*. There are eighteen members, and each member has a following of seven, often referred to as the company share holders. The contribution paid per member is N400, but this is divided between all eight people in each of the eighteen units. Registered members know their date of payment, but endeavour to conceal it from their company share holders. Often interpreted as a form of corruption, the latter do not allow their 'mediators' to escape lightly. However, this can result in conflict between share holders and his/her company members.

The other six *ojas* are based on a contribution system. Members pay their contribution on meeting days and receive their accumulated capital based on one of two systems. In some *ojas*, contribution allocations are made according to a number given to each member at the commencement of each new cycle. However in most cases, allocations depend on the leaders of the *oja*, and no one knows in advance who the beneficiary will be. There is no record of what leaders receive or gifts given to families of deceased members. Receipts are not issued for contributions, and there is no way of ensuring that the number of payments corresponds to the number of contributions collected. For example, an *oja* may have twenty-two members, but the collection of contributions could continue for twenty-four meetings. Since the leaders of the *oja* make decisions regarding payments it is very easy for them to allocate the other two contributions to 'ghost members' and pocket the money themselves. There is also an opportunity for the leaders to be bribed into helping members get their money out of turn. Each *oja* keeps a list of those who are not up to date with their contributions, and a fine is imposed. Interest is charged at the rate of 50 per cent on the outstanding balance. If a member dies in the course of the *oja* cycle his/her contribution is refunded to the family when the total contribution is completed. The family is expected to complete the outstanding payment. As soon as this is done, the family gets the full amount due to the deceased member. If a family member succeeds the deceased, there is no refund until the successor completes outstanding payments. The name of the original member is retained in the registers.

Oja leaders work on a voluntary basis, but are first in the cycle to receive their contribution and they also enjoy a donation equal to the weekly contribution from the bulk sum of a recipient on meeting days. The same man or woman can lead a number of *ojas* at the same time. Women are more numerous than men in this role, not only in Eroke itself but beyond it. A new person wishing to join an *oja*, is usually asked to 'see the leader' which implies that some token or kola is donated. For marriages or engagements, the leaders are first invited and given gifts, as their blessing is essential for other members to accept the invitation. *Ojas* are also useful meeting points for those with political aspirations, the candidate again having first to satisfy the leader. The leader would then be expected to pressurise members to vote for this person.

Even indigenous institutions have been undergoing change in the past two decades. Contributions have increased, even though not at the same rate as national inflation. Forty years ago *sisi* (sixpence) was an acceptable contribution. In ten years there has been a switch from N1 to N50 in *Lami*, while in *Juma* it jumped from N0.2 to N20. New forms of *ojas*, namely the 'share' systems of *Udama and Lami*, have been introduced and accepted even though their contribution systems are different to the regular ones. The networking advantages of *Udama* were particularly attractive, and mention has already been made about the importance of networks – particularly in a trading context.

However there are irregularities in the *ojas*, which if challenged and corrected could give the transparency and credibility so much needed for credit in Eroke. In a later section attention will be focused on how an adaptation of the *oja*, the FC project, endeavoured for many years to build up a reliable credit facility.

Credit in Eroke

Credit in Eroke is provided mainly by money lenders and the DDS. Some money lenders live within the village, but the majority reside outside. Like the situation described by Ezumah and Di Domenico (1995) for Igbo farmers in the south-east of Nigeria, neither the government nor banks offer formal credit for agriculture. Credit from DDS is largely in cash, while individual lenders may employ both cash and kind. Credit in kind is normally in the form of crop or tree produce which women market and repay the lender with interest. Trust is the main consideration in such agreements, but this cannot always be maintained. Credit given in kind is 'short-term' credit (on average for eighteen days as opposed to five months for credit as cash; Table 6.9), but the value of the goods may be high and therefore the typical amount of credit in kind is greater than cash credit. The average loan in kind is approximately N12,000 compared with N3,400 for cash (Table 6.9). Interest rates are very high, but similar for both types of credit (average annual rate of 226 to 275 per cent). Credit in cash is typically for projects of a more long-term nature, with low financial returns in the short and medium term. An example would be building or modifying a house.

Oil palm trees can be used as collateral for cash loans in Eroke, and so important is this that a whole legal system surrounds it. If a borrower cannot make the agreed refunds, the lender has the option of reporting the case to the market chief (*onu aja*), or the *gago* and *madaki* if the borrower is from the same village as the lender. The market chief only has jurisdiction over transactions undertaken in the market. In an extreme case the matter may go to court, but this is not the preferred option for all concerned. Where this happens, the lender harvests the fruits and uses them for the period of the loan – a system known as *olufa, i.e.* keeping the property in trust for money received. Witnesses are required for the making of such an agreement as it is a very serious matter to mortgage such assets, and is usually only done when a person is in dire straits. The person who mortgages his/her trees cannot harvest the trees as usual as this is now the right of the money lender. Conflict between lender and borrower is common, particularly with a resource like palm trees where yield and price may fluctuate. The borrower may approach the lender if the amount borrowed is less than the total money being realised from the palm trees, but this rarely satisfies the lender who views extra income as interest. In other instances, the family of the borrower may consider that consultation regarding the loan was not adequate, and may prevent the lender from harvesting the bunches. In such cases the intervention of the *gago* or *madaki* usually brings peace, but in extreme cases the matter may go to court. The use of tree crops as collateral for credit is common practice throughout Nigeria (Abasiekong, 1981; Adegboye, 1983).

Despite the almost complete absence of formal financial services or government initiatives, credit provision in Eroke is healthy and a positive force for sustainability. Eroke traders and farmers continue to rely heavily on credit, and living in a state of indebtedness is not a problem. Neither is it threatening to their idea of sustainability. Should these credit options ever be removed then unstainability will be declared. A good relationship exists between borrowers and lenders, and being able to avail of goods for trading on a credit basis is very important as this trade provides a lifeline for the petty trader. Richards (1990) makes an important observation about the value of such middle men, and it is doubtful whether anyone in Eroke would see them as a 'cause of agricultural stagnation' (Richards, 1990).

DDS Financial Services in Eroke

The DDS saving/loan scheme (Farmer Councils; FCs) initiated in Eroke in 1974 is a saga yielding insights and experience that led to a revamping of the system on which all its programmes were established. Figure 6.2 represents the timeline with three distinct periods of membership: mid to late 1970s, early 1980s and mid 1990s. Throughout these three periods FCs opened, closed and reformed. Members joined, withdrew, and re-entered, often to a new or different FC, and preferably in a different zone (the zones represent a grouping of the FCs with the chairman elected from one of the FCs in that zone). Loan qualification required beneficiaries to be members of an FC, and a new loan could not be issued unless the zone (the grouping of FCs) had fully repaid loan capital and interest. To avoid this problem, an FC with complete repayment could decide to change from one zone to another. This resulted in a multiplicity of zones, as members sought to avoid bad repayers. Loan repayment in Eroke had been a puzzle for many years.

The decade between the mid 1980s and mid 1990s was probably the most unsatisfactory in the history of Eroke FCs, as by 1995 only six functioned (ie. saved during that calendar year) and even these seemed moribund. However it should be noted that the early 1990s were a difficult time for most Nigerians, and compared with some neighboring villages (including other members of the 'Abo' group) the total savings in Eroke was good. The end of year balances for the period 1990 to 1993 (just prior to the onset of this study) are presented in Table 6.10. Nevertheless, given the high inflation rate in Nigeria over this period (Table 6.6), the value of the savings markedly declined in real terms. Group dynamics and leadership were important factors in the variation between the villages in Table 6.10. For example, Abo Ojuwo benefited from good co-operation and leadership, and this explains its relatively high savings between 1990 and 1993. In the experience of DDS, savings typically follow a cyclical pattern, usually declining as the members of a group age, eventually resulting in reduced savings. If this smaller group takes a loan proportionate to a fully functioning group, repayments are likely to be problematic causing diminishing financial viability and eventually closure.

Women have always played a prominent role in the Eroke FCs. Of the original twenty-five FCs, five had women members and a woman chairperson. Their savings have been consistently more stable than the men, and by 1995 they formed the bedrock of the remaining six functioning FCs. Men claim the reason for this better performance of women FCs is that they are not required to pay tax nor are they subject to the higher levies required of men. However, by the mid 1990s all were struggling to survive, and loan repayment was as bad as ever.

A major re-organisation of the FCs was initiated in Eroke in 1995 at the instigation of interested members with some encouragement from DDS. The scene improved by 1996, and women emerged with eight FCs, which they decided to arrange around their different churches so zones are now either Christian or Muslim. The reason for this, according to them, was to facilitate better repayment of loans. Other village communities have taken similar initiatives with satisfactory results, the ideas and reasons for it shared amongst other FC groups.

Because of the loan repayment record, DDS was on the point of bidding farewell to Eroke, but each time the populace at large intervened and 'begged' DDS to reconsider its decision and allow more time for the repayment of their outstanding loans. DDS agreed. Meanwhile the community took the lead in tree planting, especially NCVs of oil palms and other fruit trees. They were also among the first FCs to adopt NCVs of maize, groundnut, cassava, cowpea, rice and soybean. Their interest in nutrition was very encouraging, and they clearly wanted the very best for their children and willingly made the sacrifices required. Rainwater harvesting was another welcome innovation that enjoyed optimum community support, cash and other local contributions forming an integral part of their commitment to it.

In many respects Eroke excelled in letting its needs be known, but was secretive and reluctant in unfolding information regarding internal problems in the FC programme. Nobody was willing to betray 'community' secrets even if this secrecy undermined the viability of a whole programme. Even remote association with disloyalty can be life threatening, and very few are prepared for this sacrifice. Happily 'no condition is permanent' (to quote a common Nigerian saying), and twenty years later the community sufficiently trusted DDS to satisfy its curiosity about Eroke's poor repayments. The problems were twofold. First, the loans were not being made available at the best time for their needs, and second the loans were sometimes being used in non-productive investments making repayments impossible. An example of the latter was the use of DDS money for building new houses or replacing thatched roofs with zinc. On some occasions a sewing machine or maize mill was bought with the loan intended for agricultural inputs. There had been a slight temptation to use loans for goods such as clothes. All believed that other DDS services would be withdrawn if the above were revealed.

Once the problems were known and understood, remedial action could take place. As women are involved in post harvest activities, their interest is in having access to cash at this time. In this they are no different to their Igala sisters throughout the kingdom who like them are involved in the palm oil trade. They particularly required flexibility if they were to manage and steward the resources at their disposal. Solutions were simple and eventually came from the depths of their own vast experience, but this took time as these needed to be thoroughly thought through. Their first concern was palm oil. Hitherto, circumstances forced them to sell it prematurely, and to solve this problem a loan was required. A sum equivalent to the price they would normally receive at this unfavourable time would help overcome two problems: immediate financial needs would be met, and sale of produce postponed until optimum price could be gained. The stored palm oil served as collateral for loans taken, and refunding would be completed after the sale of the balance in storage.

A second concern revolved around cultivation of field crops. Men in the FCs normally take their loan in the months of March and April at the commencement of the agricultural year, which is appropriate for them. Women too require capital at this time, but on a smaller scale. However, it is at harvesting and marketing that women have a higher demand for capital. A loan at this time, equivalent to the market value of their processed goods, would enable them to purchase and store until prices were right. The option of a split loan, the smaller part taken in spring-time and the larger at harvest time, was in the words of the women 'the better thing' (meaning an ideal situation). An optative menu was thus initiated offering multi-choice credit to both women and men.

The above account shows a long struggle to make the DDS savings/credits meaningful to FC members. If one message is clear, it is that trust takes time to establish and if beneficiaries are to be the real target, great listening is required. This is illustrated in the fact that it took twenty years to discover what really happened to the loans. The Eroke situation was quite unique and different from the case study covered in an earlier publication (McNamara and Morse, 1998) which focused on five villages in Ibaji, (a western region of Igalaland). Despite some changes in FC membership over twenty-five years, the Ibaji FCs showed much more stability compared with Eroke, and many of the original progenitors were members twenty-five years later and still playing active leadership roles. The DDS saving/credit scheme has some very simple rules. When a community satisfactorily manages its savings and repayments, DDS was and still is happy to continue to assist with other programmes. Those most in demand are nutrition and income generation facilities. Projects of benefit to those least able to help themselves or deprived of opportunities that might benefit individuals and groups, are also essential. The secret in intervention is to strike the right balance between the two.

Credit, *ojas* and DDS – an interaction for sustainability?

As mentioned earlier, DDS's credit scheme in Eroke has been totally revamped over the past ten years. This is in the interest of building confidence within and between FC groups and zones. Members are fully aware of the regular and rigorous checking as they have participated in the discussions and group meetings that made the new rules; but this accountability makes the credit scheme less attractive to some. The *oja* leaders have certain advantages alluded to already, and some may see this as 'flexibility'. The DDS scheme is sometimes seen as inflexible because old loans have to be completely refunded before new ones can be issued, and it is not possible to bribe the management to behave otherwise. Being a member of an *oja* can also limit the funds available for the DDS scheme. For example, one FC member pays as much as N800 each week to an *oja* in Eroke. This is gradually changing as members become aware of the options now available within the DDS facility. The sustainability of the DDS system increases with each piece of new knowledge, and its efficiency and flexibility is contributing to members confidence in it. The statements of accounts issued to each FC, twice yearly, and the participatory methods used to encourage accountability by members should enable them to think critically about the programme as a whole. The long-term dream would be the enhancement of the *oja* system by the inclusion of some of the practises that have made the DDS scheme transparent.

The decision of some Eroke FC members to divert their loan fund to house building also provides food for thought. In hindsight it was a wise decision, as the defaulters would otherwise perhaps never have been able to afford a house of their own, especially as building materials were available in abundance and at keen prices in the late 1970s and early 1980s. However, as the money had not been invested in an activity that generates revenue then refunding became difficult. If DDS had known the truth earlier then remedial action could have been jointly planned, but since fear is the first stage of wisdom, this phase can be put down to experience. The greater the range of choices the better the prospects for sustainability, but this only comes with experience and maturity.

How long does it take to establish trust? It may seem to have taken longer than normal in Eroke, but credit appears to be a very sensitive topic. Now that a re-organisation has been successfully concluded, with a range of different and more versatile options open to members, the future looks more promising. The initiatives taken by the women show they value it, otherwise they would not have bothered to find ways of managing and controlling it more efficiently and effectively within the auspices of their various churches. Trust will always be problematic, but the position has decidedly improved as information is shared and values internalised.

There will always be reasons for destabilising group membership, but when there is sufficient openness among members, threats can be treated as challenges. Some unhealthy alliances were noticed among a few members who sought to exclude potential members from joining the FCs. Their incorrect understanding was that it would limit the loan fund available to individual councils. If outsiders like DDS can introduce delicate topics that are potentially threatening to communities, discussion is facilitated as blame is not apportioned to an individual or group. The need to ensure that old and new members are aware of privileges and conditions is of paramount importance. During DDS evaluations of its credit programmes some Eroke members alleged they were not aware it was possible to borrow their own savings. They believed that interest paid by them on their savings went to DDS, when in fact no interest is charged when members take their savings as loans.

Some petty traders still continue with their age-old system of buying grains on credit and refunding two weeks later in cash. A trader who gets N400 worth of goods on credit may have to pay an annual interest rate of 250 per cent, (if not more) and refund N440 two weeks later once the goods are sold. While this has many positive aspects, it may be argued that such a system increases the cost of food to consumers. The transition to a less costly method is difficult mainly because they do not believe that it is possible, and they do not wish to gamble with something untried. However with a newer and more complete understanding of the saving and loan scheme this is more likely to change.

Leisure

When it comes to leisure, Eroke inhabitants look to the book of Genesis which describes how God looked at his own creation and saw that it was good and so on the seventh day he rested. There are no details given, but suffice it to say the Creator took time out to enjoy and admire the work of his or her hands and to recover from the hard work involved. Rest and leisure are integral to life.

In the past, explorers and traders understood the riches of Africa and had no compunction about trading valuable goods such as gold, ivory, ebony and hard woods at a poor rate of exchange. The list of the resources taken away from Africa is endless, but some things were elusive and therefore escaped the greedy touch of the invaders. Leisure is certainly in this category, but institutions such as the *oja* and labour groups also remained unscathed. Chapter 3 describes how Eroke has the innate ability to relax and enjoy life. Rests are routine after the toil of the mornings and before returning to evening duties. Further relaxation is integral to evening activities at household and compound level. The official rest days on Fridays and Sundays show this extended to the community as a whole.

However, it is during the time of the new moon that leisure and celebration take on real momentum. This is the time set especially aside for music, song, dance and feasting. Adjacent compounds join the celebration with young and old all participating. Eroke is renowned for its music especially on wind and string instruments and its particular type of drumming. Singing is also popular, and Eroke choirs enter many competitions. Repertoires have greatly increased over the years and so have performance skills. Great care is taken to ensure that youths and children are initiated into all aspects of these arts. These moonlight nights provide the ideal setting for these to be inextricably woven into the very fabric of their beings. All eventually sing, dance and move in that perfect rhythm leading to a high crescendo which transports participants into that domain where it is possible to forget cares and worries. It is a therapeutic exercise, out of which evolves courage and enthusiasm to take on the world afresh. Story telling and recounting parables and fables are also important, and provide 'moral' teachings drawn upon in times of decision making and crises. They help to keep alive the history of Eroke as do many of their dances. Merrymaking revolves around palm wine drinking. These lunar activities energise and renew the village communities.

There are however other by-products from their entertainment and leisure. Because many of the troupes are so expert and professional, they are hired to entertain guests at weddings, wakes and funerals. Fees are negotiable, but N5,000 to N10,000 can be earned for a few hours performance. They also enjoy the prestige attached, and of course the prospects of new contacts is a big appeal.

What role does leisure play in sustainability? This is an almost impossible question to answer. Leisure is therapeutic and entertaining as well as a low cost and enjoyable system of transferring history and skills from one generation to the next. It is certainly true that leisure is rarely, if ever, included in SI lists beyond an assumed pooling with vaguely defined notions of 'quality of life'. Even more holistic umbrella terms like 'sustainable livelihoods' or even sustainable development, as opposed to perhaps more narrow concepts of sustainable agriculture, hardly ever include a consideration of leisure. Leisure is difficult if not impossible to quantify, and is influenced by a host of factors such as the macro-economic picture (Lawuyi and Falola, 1992), yet it is an important if not vital facet of life. Livelihood and leisure interact and intertwine to such an extent that one cannot exist without the other. Other have noted its importance, and the following quote from Guyer referring to male smallholders in south-west Nigeria is an example:

> the leisure of these male smallholders is by no means residual; it may, in fact, help to determine the pattern of work, by comprising activities through which production is given meaning or made possible at all.

Guyer (1992)

Leisure in Eroke is strong, rich and vibrant, and certainly provides a positive contribution to sustainability.

Self Reliance

The inhabitants of Eroke take pride in their history of self-reliance. However only particular aspects of this heritage are selected here and only in so far as they are pertinent to the sustainability debate. So self-reliant was Eroke once, that it only imported one commodity namely salt. Even then it could produce it if the need arose, and to this day there are women who still know how to make it. It is not surprising that much of their self-reliance can be attributed to the abundance of oil palm, but it should also be noted that the population was much smaller in the past. Clothing and shoes were made from trees and cotton, employing skills that all women knew. There was no shortage of labour as most young men stayed at home. Children were many, though of course they admit to some dying at birth or at a young age as no inoculations were available. Some women died during child birth, although one of the oral historians claimed that such complications 'were brought by the Europeans' and their 'modern medicine'. Eroke was well known for its herbal treatments and bone setting, and although this tradition still thrives many of the people who practice nowadays are accredited to schools of traditional medicine in cities like Kaduna.

The first erosion of self-reliance came with the importation of cloth, aluminium pots and pans, iron beds and corrugated iron sheets for roofing. The latter being safer than the traditional use of thatch afforded protection against fire. In the words of oral historians, Eroke 'became a safer place' at the same time as its self-reliance was changing its direction. Safety was further enhanced by the reduction of wild animals, as hunters were plentiful in the past. Money earned from hunting was invested in the new roofing materials, and locally made traps, bows and arrows and catapults were replaced by better traps, 'Dane' guns and eventually the modern rifle. Locally made gunpowder and 'cap' were used in the 'Dane' gun, but because both of these became useless in the rain the cartridge was a very welcome innovation. In due course the population of both big and small animals gradually dwindled until the current restrictions prohibiting hunting were introduced. Imported cloth reduced the workload of women though it destroyed the indigenous industry. These 'imported' goods rapidly became symbols of prosperity, and those who were better off were clearly recognisable as they lived in a corrugated iron roofed house or rode a bicycle. The negative side was that it highlighted social differentiation, something not so obvious until then.

Migration is a factor that impacted all Igala as from the 1950s. As the world price for coffee and cocoa increased, large numbers of Igala men and their wives, mostly illiterate, migrated to the coffee and cocoa producing areas in the west of Nigeria or even further afield to work as labourers. Many, if not the majority, of these were from Imane, Ogugu and Okpo districts all near Ankpa. Most of them did not return, but the more ambitious among them succeeded in buying and managing their own cocoa farms and became rich as a result. There was also migration to the north of Nigeria during colonial days, especially when Kaduna was the capital of the northern region. Some succeeded in acquiring white collared jobs. The development of textile industries in the north in the early 1960s also provided a source of employment. Another set of Igalas who migrated to the north were young Moslems who enrolled as students with Moslem scholars, and again most of these were from the Ankpa area including Imane. The tradition of migration, especially to the west, continues to this day, and Igalas can be found all over Nigeria. There are sizeable Igala populations in most Nigerian cities, including Lagos, Abuja, Kano and Ibadan. As the migrants continued to maintain strong links with their home the flow of information proved invaluable in terms of long-distance trade. Remittances also became very important to the village economy.

Education was first introduced in Eroke in the mid 1940s. Initially it was rejected as farming was the only occupation that mattered. Nearby villages were 'more enlightened' according to the female oral historians in Eroke, and these sent their boys to school. Eroke women in particular, seriously regret not being sent to school, and still bemoan the decision taken by parents on their behalf. Because of this, nothing will now deter their determination to change this position, and they work hard to ensure their children, boys and girls, attend school. However, education impacts on migration as once educated the young men and women wish to leave the village and find employment outside - most do not return.

> A large majority of Africans still intend to return home, but many never make it.
>
> Peil *et al.* (1988)

It is perhaps ironical that in other parts of Nigeria building schools has been seen as a mechanism for keeping a population (Stone, 1998). In Eroke, and one suspects many other places in Nigeria, the effect has been the opposite.

Eroke women in particular have a great ability to analyse their situation. They want change and are currently paying the price for it as most of their energy is deployed to making enough money to cover the school fees of their children. This has accelerated the rate of emigration, especially the educated young men and women, leaving them with a great burden of work. There is as yet no compromise by these women. Will this ultimately lead to destabilisation of livelihoods for who will succeed them, and even if some do will they work and strive for similar aspirations? New information and perhaps change at the macro level will continue to influence Eroke in the future as it did in the past.

Quality of life and sustainability

Many of the indications suggest that Eroke livelihoods are sustainable. There are many opportunities for them to generate income, and they are flexible and adept at making use of these. Access to financial services of one form or another is good, despite the absence of government initiatives and the relative paucity of formal bank services. Although there are complaints about high interest rates, no one regards the local credit services as exploitative – its just business. Similar views held by Richards (1990) regarding the role of money lenders in Mogbuama zone of Sierra Leone resonate well with findings in Eroke. Problems of varying degrees undeniably exist but not all are threatening to sustainability. Some people are disadvantaged due to circumstances of ill health and bad situations not of their making.

Quality of life is a subjective and nebulous term, and like the bigger umbrella of sustainability it is heavily value-laden. At one level it is simply equated to net income (income left after all necessary payments have been made), but this is deceptive. In the 1997/98 production surveys the number of respondents who believed that personal and household net income had decreased is matched by the number who claimed an increase (Table 6.11). Few thought income had remained the same, and change in some direction was almost unanimous. When asked whether others were faring better than before (i.e. had the income distribution in the village improved or worsened?), the majority cited a worsening situation claiming the better off had become much richer. The gap between rich and poor may be wider, but the quality of life was better for everybody than ten years previously. Nobody perceives themselves to be poor as this would be an abomination for the family concerned.

Another indicator of quality of life that is often used is health, and in Eroke there was no evidence of a decline although there are exceptions. Those claiming a worsening situation are more than matched by those who

Livelihood and Leisure 195

say things are improving (Table 6.11). What one is seeing with both income and health is probably a variety of opinions largely linked to individual circumstance rather than widespread distress. The final caveat in this discussion is that a clear majority felt their quality of life and 'enjoyment' had improved over the last ten years (Table 6.11).

What can be concluded from this analysis of livelihood and leisure with regard to sustainability? A summary of the main points is shown in Table 6.12. There is a diverse range of occupation and sources of income, with no monopolistic reliance on crop production although it is clearly important. People have the option of spreading income over a number of sources, and are able to switch between them as opportunities arise. The importance of off-farm income in African agriculture is well-known and documented, and investment is typically said to flow to the former rather than the latter (Flinn and Zuckerman, 1981). Change in net income was largely a matter of personal circumstance, but no evidence of 'village wide' problems. There was more unanimity about income distribution – the consensus being that the rich had got a lot richer! Expenditure is less flexible, especially in education.

In health there are some options as herbalists provide an alternative to expensive 'western' medicine and the DDS clinic/demonstration advise on nutrition. Given the extensive networks that operate, one could also argue that some flexibility exists with purchases of food. Credit provision is generally good, although interest rates are high. There are different types of credit provision, long and short term adapted to meet the needs of both producers and customers. Local institutions such as the *oja* clearly play a major role, including an enhancement of networking, with variation between the various types of *oja* to suit needs. Leisure is vibrant. All of these point strongly towards sustainability, but there are problems. Outside intervention like the DDS programmes took time and energy before a good understanding was established. This was eventually achieved due in part to this study and to the reflection – action process. Education is acting as a passport out of Eroke and an increasing number of young people are making use of it. On the down side, this exodus will leave the village devoid of an adult population.

This chapter provides a further interesting point for discussion with regard to spatial considerations of agricultural sustainability. At the onset of this research it was decided to focus on a village, a choice confirmed later by Izac and Swift (1994). Spatial scales, although tempered by awareness of flows across boundaries, sit at the heart of much of the literature on gauging agricultural sustainability or indeed sustainable development. A tangible entity that can be defined on a map is clearly an attractive base from which to start any analysis – especially something as complex as sustainability. Yet there is much irony here. After all, the porosity of the village boundary in terms of income and the importance of external market forces as initiators of agricultural change have been well known for long. Clarke (1981) and Goldman and Smith (1995) cite such factors as being critical in agricultural transformation. Given that markets are often located far from the production base, does it make sense to use a village as the spatial unit to examine sustainability? This question will be addressed in more depth later, but if many of the forces for change originate outside the village, then clearly the village in a primary physical and spatial sense can potentially be very limiting.

To summarise, livelihoods in Eroke have an innate resilience that can only be positive for sustainability. Given the diversity and internal/external pressures, stasis is not to be expected. Change is the norm, and if anything the turbulence and uncertainty within the country has enhanced flexibility rather than destroyed it. Any threat to sustainability would be due to circumstances that did not allow Eroke to adopt/adapt on time. The secret is the availability of options and the ability to use them. Again, this is not new and a host of findings from others emphasise the importance of diversity and adaptability, even if disagreeing somewhat on details.

Table 6.1 **Number of occupations and sources of income per respondent**

Category	Number	Male	Female	Chi-square and F-test
Occupations	1 to 2	11 (1.46)	18 (-1.46)	ns (1)
	3 and greater	14 (-1.46)	33 (1.46)	
	N	25	51	
Sources of income	1 to 3	9 (2.75)	10 (-2.75)	ns (1)
	4 and greater	16 (-2.75)	41 (2.75)	
	N	25	51	
Mean age (N; SD)		49.2 (24; 16.8)	31.1 (49; 13.8)	*** error df = 71

N = number of respondents (survey conducted in 1997)
SD = standard deviation
ns = not significant at the 5% probability level; *** $P < 0.001$

Table 6.2 Types of trader and goods traded in Eroke

(a) Type of trader

Category	Male	Female
petty trading	2	5
short-distance trader		9
long-distance trader		18

(b) Goods traded

Category	Goods	Male	Female
Food	processed food	1	16
	unprocessed food		16
Food ingredients	small quantities		11
	wholesale	2	16
Others	artifacts	2	
	animals	2	
	clothes		1

Twenty respondents (2 male and 18 female) known to be traders. Each clan was represented in the survey, and an individual may list a number of categories as applying to themselves.

Table 6.3 Type of remittance sent back to Eroke by 'abroadians'

Respondent category	N	Items sent to Eroke by 'abroadians'			
		Clothes etc.	money	Food/drink	Others
Household head	11	11	10	3	5
Mother	10	10	9	4	1
Sisters	8	8	7	2	1
Wife(s)	4	4	1	0	0
Other relations	6	1	6	4	2
Friends	9	2	1	9	1
Totals	48	36	34	22	10

Number of respondents (N) = 48 (taken from 20 households). Figures represent the number of people in that category who mentioned receiving that item as a remittance from an 'abroadian' within the previous year (1995 to 1996).

Table 6.4 Skills and other agricultural inputs brought back by migrants

Contribution from migrants	Yes	No
New knowledge and skills	10	4
Seeds, planting material, agro-inputs	7	6

Number of respondents = 14 (all male, who had been in contact with a returning migrant). Figures represent the number of respondents who had received that particular contribution from a returning migrant. Survey conducted in 1996.

Table 6.5 Inflation in education and health care costs (1988 to 1998)

(a) Education items and charges

School	Percentage increase (1988 – 1998)
Local Government School, Efofe-Imane	2324
Abo-Ojache Primary School	1206
Abo-Ojikpadala Primary School	345
Winners Primary School, Imane	2913
Roman Catholic Primary School, Imane	667
Local Government Central School, Imane	486
Imane Community Grammer School	421
St. Charles School, Ankpa	379
Mean increase (SD for schools)	1093 (994)

(b) health care items and charges

Clinic	Percentage increase (1988 – 1998)
Health Clinic, Ofudu-Imane	74
Udama Dispensary, Imane	1167
Rapheka Clinic, Imane	395
Mean increase (SD for clinics)	545 (562)

Official consumer price inflation (1988 - 1995) = **1026%** (source : IMF Financial Statistics Yearbook)

SD = standard deviation

Table 6.6 General inflation in Eroke (1970 to 1980 and 1980 to 1996)

	1970 - 1980		1980 - 1996	
Item	N	Mean	N	Mean
cutlass	9	1379	9	1300
bucket	9	188	9	1293
hoe	5	385	5	1280
kerosene (4 litres)	12	520	12	6399
petrol (4 litres)	4	124	4	1990
maize (sack)	11	183	12	5715
rice (sack)	12	94	12	6378
yam tubers (3)	13	907	13	2233
cassava (basin)	13	3073	13	3038
slippers	14	533	15	5106
hen	15	2887	15	11425
vegetable oil (bottle)	4	1033	5	2198
palm oil (bottle)	11	765	12	9830
bread (loaf)	15	492	15	5324
soap (low quality)	7	571	10	7272
soap (high quality)	10	337	12	11237
washing powder (packet)	13	400	15	9510
baby expenses	9	416	9	4451
Xmas cloth	14	505	15	5890
zinc roof panels (bundle)	4	159	4	7352
cement (bag)	4	350	4	4529
taxi fare	7	593	9	1996
bicycle	3	252	3	7400
copy book (school)	7	488	7	10126
pencil	7	1074	7	4412
Overall mean		708		5507
SE		754		3195

Official consumer price inflation: 1970 - 1980 = **306%**
(source : IMF Financial Statistics Yearbook)   1980 - 1995 = **4804%**

N = number of respondents for that item
Mean = mean percentage increase over the period in question
SE = standard error

Table 6.7 The number of visits in two years (1993 to 1995) to three sources of medical treatment/help

Number of respondents = 21 families (survey conducted in 1995)
Numbers in parentheses are deviation from expected values.

Age group	Sex	Herbalists	Clinic/ hospital	DDS clinic/ demonstration	Sig. (df)
Elder	M	31 (5.27)	17 (-4.33)	16 (-0.94)	* (2)
	F	10 (-5.27)	17 (4.33)	11 (0.94)	
Adult	M	75 (-3.39)	118 (16.95)	68 (-13.56)	** (2)
	F	98 (3.39)	105 (-16.95)	112 (13.56)	
Children	M	40 (7.17)	65 (-1.12)	92 (-6.05)	ns (2)
	F	33 (-7.17)	82 (1.12)	126 (6.05)	
Totals		287	404	425	

M = male
F = female
ns = not significant at the 5% probability level; * $P < 0.05$; ** $P < 0.01$

Table 6.8 Some of the characteristics of the eight Eroke *ojas*

Oja	Meaning	Period	Member.	Fees	System
Udama	unity (Igala)	9	both	yes	share
Lami (2 ojas)	Thursday (Hausa)	8 and 15	both	no	contrib. and share
St Cecilia	Catholic Church based	8	women	yes	contrib.
Juma	Friday (Hausa)	8	both	no	contrib.
Ladi	Sunday (Hausa)	8	both	no	contrib.
Afor	*Afor* is a market day	13	both	no	contrib.
Igeya	same as *Afor*	13	both	no	contrib.

period	days between meetings
fees	payment of membership fees
member.	membership (both = men and women)
contrib.	contribution

Table 6.9 Comparison of credit taken as cash with credit taken as kind

Statistic	Credit in cash			Credit in kind		
	Amount	Months	Rate	Amount	Months	Rate
N	33	30	33	40	40	40
Max.	32,000	12	2,400	80,000	3	2,000
Min.	400	0.25	0	1,000	0.25	0
Mean	3,397 **	5.19 ***	226.1 ns	12,257 **	0.62 ***	274.9 ns
df	1, 71	1, 68	1, 71	1, 71	1, 68	1, 71
SD	5,695	4.57	500.5	15,120	0.44	362.5

N   number of respondents
SD   standard deviation
Amount amount of credit taken (Naira)
Months number of months over which loan was taken
Rate   annual interest rate (%)

ANOVA results (one way) refer to credit as cash compared with credit as kind. ns = not significant at 5%; ** $P < 0.01$; *** $P < 0.001$

Table 6.10 Comparison of the end of year balances in Eroke with some neighbouring villages (1990 to 1993)

Village	end of year balance (Naira)				Difference (90 to 93) in balance
	1990	1991	1992	1993	
Eroke	1,805	1,836	2,315	2,315	510
1	847	916	127	270	-577
2	3,815	3,108	6,228	3,656	-159
3 (Abo Ojuwo)	1,877	1,388	1,877	2,479	602
4	1,261	1,123	1,443	1,552	291
5	911	972	1,189	1,270	359
6	2,329	2,276	2,867	3,078	749
7	0	0	680	980	980

Villages 1 to 7 are neighbours of Eroke (mostly in the 'Abo' group).

Table 6.11 Perceived change in quality of life in Eroke (late 1980s to late 1990s)

Indicator	Increase (better)	Same	Decrease (worse)	Sig. (df)
Personal income	31(-3.61)	5 (1.09)	35 (2.53)	ns (4)
Household income	31 (-3.61)	8 (-1.91)	32 (5.53)	$P = 0.06$
Income distribution	19 (7.23)	5 (0.83)	44 (-8.06)	
Health of household	35 (2.26)	9 (0.57)	27 (-2.83)	ns (2)
Health of children	39 (-2.26)	10 (-0.57)	21 (2.83)	
Quality of life	43	11	17	

ns = not significant at the 5% probability level

Table 6.12 A summary of livelihood and leisure indicators in terms of sustainability

Area of concern	Summary	Some indicators
Occupations	sustainable	diverse range of occupations ability to identify new occupations and adapt
Income	sustainable	diverse range of income sources good ability to adapt and switch between sources individual and household net income has broadly remained constant (varies between individuals) income distribution has worsened
Expenditure	sustainable	can readily import shortfalls health and education have become more expensive, but costs have risen at a rate less than inflation range of options for health care
Credit facilities	sustainable	flexible range of credit facilities interest rates are high
Oja meetings	sustainable (but?)	flexible range of *oja* meetings open to corruption and misuse
DDS FC project	unsustainable but moving towards sustainable	history of turbulence many changes in membership poor savings and repayment of loans history of transparency and trust
Leisure	sustainable	strong and vibrant sources of leisure
Self-reliance	unsustainable	self-reliance has declined emphasis on education encourages migration of young out of Eroke
Quality of life	sustainable	health of household and children has remained much the same quality of life is said to have improved

(a) Male income

Trading	Farming	Other	Tree crops	Remittances	Animals
1.5	2	2	3	4	4

(Underline groups: Trading–Farming–Other–Tree crops; Remittances–Animals)

(b) Female income

Trading	Other	Farming	Tree crops	Remittances	Animals
1	1	3	3	4	4

(Underline groups: Trading–Other; Farming–Tree crops; Remittances–Animals)

(c) Male expenditure

Food	Health	School fees	Farming	Clothes	Funerals	Transport	Water
2	2	3	4	5	5	7	8

(Underline groups: Food–Health–School fees; Farming–Clothes; Clothes–Funerals)

(d) Female expenditure

Health	Food	School fees	Clothes	Funerals	Farming	Transport	Water
2	2	3	3	4.5	5	6	7

(Underline groups: Health–Food; Food–School fees–Clothes; Funerals–Farming)

The lines joining different categories indicate that they are not significantly different at the 5% probability level (details in Appendix C). The figures are the median ranks for that category, with a rank of 1 being the highest.

Figure 6.1 Comparison of median ranks of male and female income and expenditure

Year(s)	Events
74	DDS employs an extension agent in Eroke
79	25 FC's started in Eroke
80	Zonal farm started (just over 1 ha)
	Extension agent leaves in early 1980 for further training (1 year) and hands over to a colleague
	Extension agent returns at the end of 1980 to find only 5 FC's remaining (4 of them chaired by members of the extension agents family)
82	9 new FC's started (some new members and some made up from members of the original 25 FC's).
	The 5 FC's remaining from 1980 were called Zone I and the new FC's were called Zone II
82-95	Much turbulence in the FC's. Many leave and some return. Major problem was loan repayment. By 1995 only 6 functioning FC's remained from both Zones I and II, and these were mostly female (5 FC's). Christian and moslem women were mixed within the FC's
95	Creation of Eroke Zone III. Originally 11 FC's but 2 quickly closed leaving 9. At this point the total number of functioning FC's in Eroke = 13 (Zones I, II and III)
	8 of the new FC's comprising Zone III were women's groups, but Christian and Muslim women were separate (5 Christian and 3 Muslim)
96	Much confusion began to arise as to the numbering of FC's in Eroke because of the 3 zones and continuing volatility of membership. Decision to re-number the FC's as 1 zone.
97	A total of 18 functioning FC's exist in Eroke, although level of activity for many is very low.

Figure 6.2 Eroke Farmer Councils (FC's) timeline

7 Visions of Sustainability

Introduction

As seen in the previous chapters, the notion of defining and measuring agricultural sustainability in Eroke is problematic. The demographic changes discussed in Chapters 3 and 4 pointed to unsustainability, for a village that has lost its young men and women can hardly be sustainable. At first, Chapter 5 seems to reinforce this message as depleted soil resulted in decreased production and various strategies designed to cope with this, including the planting of more legumes. However the message became more complicated as consumption patterns did not indicate a problem. Labour was also being employed in a wide range of activities other than crop production. Chapter 6 on livelihoods threw the situation back into doubt as there was no obvious situation of unsustainability in the village. It appeared as if unsustainability in certain areas was compensated by sustainability in others – an internal correction.

This chapter explores the ambiguity further, and draws some conclusions by examining sustainability from a western perspective. It looks at the suitability of the methodology employed. The work took a long time to complete, and therefore two questions might be asked – was it worth the time and effort, and are there alternatives? Finally, did the local vision of sustainability in any way match the western perspective and whose vision counts?

Spatial and Time Frames

Setting of appropriate scales for analysing sustainability receives much attention in the literature. Spatial scales, with their geographic and agro-ecological units provide boundaries within which one can operate, though it is clear that these are permeable. Using the village catchment in this study as rationalised by Izac and Swift (1995) provided a logical starting point. It was possible to look at demographic change, production and broad trends in labour, soil fertility management, pest and diseases and availability of inputs within the confines of a 'village catchment'. However, it soon became apparent that this model could be constraining. The carrying capacity of the village was propelled far beyond physical limits to its people scattered in all parts of the federation. Chapter 6 projects a picture of an emerging vibrant economy based on trade and services, with agriculture playing a subsistence role. Foodstuffs are increasingly being imported into the village, making spatial boundaries permeable and almost irrelevant. If enough money is earned from trade, especially in tree crop produce, to allow its people to purchase food in shortfall, why can't Eroke be sustainable? A narrow focus on agricultural sustainability may trigger alarm bells, but many admit they are better off now than they were ten years previously.

Most would accept the permeability of boundaries and criticise an attempt to make them inflexible. The problem with inventing physical boundaries is that they create a wrong impression and engender a myopic vision of sustainability. It would be impossible to disengage sustainability from the people of Eroke as they are the actors engaged in the activities being studied. A social rather than a physical orientation may minimise this danger as Nigerian villages are primarily social constructs with a physical location. The danger of the 'bounded village' as a unit for analysis has been discussed (see Guyer 1997 for a brief summary).

A reasoned argument was made for the timescale chosen (late 1980s to late 1990s), and the beginning of SAP was a logical starting point. Others have reasoned that this timescale is a realistic minimum for an assessment (Izac and Swift, 1994), but the problem with any timescale is how to extrapolate for the future. The question is whether their current sources of income will be enough to sustain their livelihoods for the next ten years or more. There are no indications that these would or would not continue, but all the evidence bespeaks a remarkable resilience within the population due to their ability to access a wide range of resources and creativity in employing them regardless of instabilities at national level. Local visions may be short-term, but they are in line with what people in Eroke think about sustainability.

Methodology

The research was sometimes tedious as it involved years of formal and informal data collection. Even with the advantages of local knowledge, languages and long association with Eroke, triangulation was always critical and eventually rewarding. A top-down approach was unavoidable given the need to build upon the frameworks and ideas summarised in Chapter 1, making it necessary to explain every detail a number of times. Sustainability is a creation of the western world, and does not easily allow a meeting of minds. Consequently some of the concepts didn't always resonate with local understanding.

Another consideration was the time it took to carry out the work. It took six years to complete, and the first four years were devoted mainly to intensive qualitative data collection, for a couple of weeks every month, together with some quantitative work. It was a punctuated exercise, giving the authors time to analyse and interpret results before feeding them back to the respondents as an iterative exercise. However, even with the advantages of prior experience, knowledge and trust, there were some shortcomings as respondents presumed the authors to have more information than they actually did. The last two years were geared to the quantitative and the generation of numbers that provides statistical proof of what was seen, rather than rely solely on the qualitative.

Some of the tools of RRA and PRA were employed in this research, though it was not primarily an exercise in either; its purpose was to get information and not to catalyse change. An interesting speculation would be whether RRA or PRA could have generated the same result? An RRA would certainly be much cheaper, but what would be detected? Much would depend on the expertise of the team, but even a cursory glance by a seasoned eye would detect the following:

1) Poor soil and yields especially of maize. Coping strategies are clearly visible, most notably in the increased cultivation of legumes.

2) Aged and aging male population. Basic questions would elucidate quickly the extent of emigration of young people from the village, particularly as this is a well-established feature of other Nigerian villages (Guyer, 1997) and long evident in the country (Swindell, 1984).

The advantage of a longer period allowed the researchers to distance themselves from the topic and return to it with a new perspective, especially following discussion with key resource persons. The more subtle points that make for sustainability are likely to be missed in RRA. The contentious issue of women owning tree crops took years of iteration and triangulation as one came to terms with the transformation in progress. Could an RRA team pick this up?

The six year span also allowed DDS to do an ongoing evaluation in a village where it had worked for twenty years. There was consensus in Eroke regarding the need for ongoing evaluation and they invited the researchers to come and 'live their lives'. Results could be applied in many surrounding villages with similar DDS operations.

The sustainability of Eroke – their own view

Sustainability has no counterpart word in the Igala language, though many of the indicators listed in Chapter 1 would resonate with what exists in Eroke. There is an abundance of local knowledge regarding soils and crops, and problems of poor fertility are well appreciated. Eroke is still 'home' to many, especially for those living abroad in towns and cities. They live for the present, in all its profundity, laying solid foundations for the future as they perceive it. One respondent when asked whether he would like his sons ('abroadians') to return said 'yes – very much'. Yet when reminded of an earlier comment regarding land shortage, his response was:

> that's fine – I'll be dead by the time this could become a problem if the abroadians return, so no problem!

In the western sense, self-reliance is distinct from sustainability (Manyong and Degand, 1997). However, distinguishing sustainability from self-reliance became something of a problem for the respondents. A tradition of self-reliance has always existed in Eroke amongst both men and women. So self-reliant were they that in the living memory of some participants the only commodity they could remember importing was salt, and given a scarcity they could have produced it themselves. As mentioned in Chapter 2, the first erosion of this independence was the importation of tin roofing sheets, iron beds, aluminium pots and pans. These were welcome because of their durability. Thatched roofs are fire hazards and bamboo beds were considered to be unhygienic, clay pots were easily broken. The introduction of these commodities made life easier for men and women, but marked the commencement of dependence on imports.

During a meeting of oral historians, self-help seemed to be a central theme, and the decision was taken that efforts be made to retain the indigenous skills now almost extinct. Candle making from palm residue was revived and much appreciated due to the high cost and scarcity of kerosene. There is still evidence of an internal reserve upon which they can draw when faced with increasing import prices, but this has its limits, and it is noted how the village has moved away from self-reliance in food. Their ability to survive now depends largely on interaction, especially in trade, with other villages, towns and cities.

The family, including those abroad, is the singular most important part of life, and mothers especially work ceaselessly for its continued sustenance and improvement. Engaging social units and livelihoods makes agriculture one of the key elements for consideration, but any concept of sustainability must ultimately emanate from what communities cherish most.

DDS development in Eroke: has it facilitated sustainability?

DDS has been involved in development in Eroke since the mid-1970s, and has always sought to combine the technical and social dimensions in its programme. Trust on the part of DDS and its intended beneficiaries have been a vital consideration throughout. Some initiatives can be regarded as sustainable while others have proved to be less so. A good example of the latter is the savings and credit scheme. This was unstable for many years until the reasons giving rise to its problems were eventually unearthed. Now the situation has changed, but not before three different phases were experienced (Chapter 6). If the current position can continue, the scene is set for transparency and accountability that should in theory make for a more sustainable situation. Adaptations in the loan scheme to suit seasonal requirements are in place, but are they important as most farmers and traders are attuned to live in a state of perpetual indebtedness? Ideally the local *ojas*, the institution on which the DDS savings/credit scheme is founded, could benefit from the recent upgrading of the credit scheme. However, it is difficult to say for certain that this will help as flexibility is reduced when it is no longer possible 'to bribe management'. More openness could solve this problem as ideally development intervention needs some freedom if intended beneficiaries are to really benefit. Otherwise, is it sustainable?

The DDS water scheme is sustainable and built to last for at least fifteen to twenty years provided ordinary maintenance is carried out. Health care has changed as traditional herbal treatments are now back in vogue. Because of their experience with both western and traditional treatments there is now a more thorough understanding of how these work. A clear distinction was always possible between sickness to be treated with either traditional or western medicine. As practitioners are now accredited and well aware of the much consultation between the two branches, a happy state of inter-dependence is being established. Improved nutrition and good quality water should also help. DDS encourages this co-operation and sees it as a contribution to stability and sustainability with a remit far wider than Eroke. DDS has been active in responding to requests for NCVs of both field and tree crops. Cassava NCVs seem to have made the most striking contribution to sustainability because of their capacity to yield well even on the poor soils, resistance to pests and diseases and more importantly, similar consistency and taste to yam.

Although involved in tree crops almost by default, this intervention has helped develop the oil palm industry to its present high level. DDS has therefore, assisted Eroke pursue aspects of its livelihoods that have repercussions for sustainability by their own definition as well as expressed needs.

Is agriculture in Eroke sustainable?

Area and yields of various crops in Eroke are in a state of flux, and there is a dynamic pattern of crop variety testing and selection. Farmer's actions depend to an extent on the resources available, but there are many other considerations. Does change *per se* amount to unsustainability? Change can occur for a variety of reasons, and the story of the Kofyar on the Benue Plains in Nigeria is a good example (Netting *et al.*, 1989; Stone *et al.*, 1990, 1995; Netting and Stone, 1996; Stone, 1998). This group farmed intensively on the highland soils in Plateau state, but gradually moved to the more fertile lowland soils along the edge of the River Benue. Initially they practiced extensive agriculture, but gradually settled back to their familiar intensive form. The major influence upon what the Kofyar grew and how was the market, although change was also catalysed by contact with other groups (such as the Tiv) then beginning to farm in the same area. The Boserup (1965) concept of population and market pressure on resources forcing change is but one model. Other factors include time management to guarantee leisure activities (Guyer, 1992).

It is true that some of the change in Eroke has been driven by the poor nature of the soils allied to an ageing population. Continually working land near compounds puts pressure on the soil, and cropping intensities have remained high even if much depends upon individual circumstance. Limited coping strategies have been put into place, notably an increase in legume planting and a decrease in cereals, but none of these are drastic. An increase in cassava planting can be explained in ways other than a response to worsening soils. A snapshot of this may suggest unsustainability to a natural scientist, but at the same time there are no signs of imminent collapse, as farmers manage a response to each new circumstance. Land is not being carved into smaller parcels because abroadians are not returning, and some residents still have access to large tracts of land. It is reasonable to predict that the present exodus will continue, and therefore pressure on land is likely to remain the same or could even decrease. Land close to compounds will continue to be heavily farmed and old fallow brought back into use for high value crops when required. The final result may be termed agricultural sustainability. But does it then follow that Eroke is sustainable?

If sustainable implies constancy and permanence, the answer is no. If the young continue to leave, households will gradually dwindle. Out of the sample of twenty households, it is highly likely that only one will be fully viable in another fifteen to twenty years. Therefore it is probable that Eroke will continue in a new and smaller way, but does this mean unsustainability? If it does, the following quotation (Goldman, 1995) holds true:

> There is no a priori reason – and certainly little likelihood - that everything should be sustainable.

The key to survival will be the availability of options and an ability to creatively adapt them in each particular situation. This is sustainability but not preservation. The research set out to answer a specific question – can the sustainability of agriculture be measured? The key result to emerge from this investigation is that the original question was curtailing and begged for a more inclusive approach capturing the real essence of life in Eroke.

Measuring sustainability – did SIs help?

For all that has been said in the preceding section, one can reasonably ask – did the use of SIs help the researchers arrive at some measurement of agricultural sustainability in Eroke? It has already been argued that a focus on agricultural sustainability provides the basis of only a very limited and incomplete analysis, so how could SIs make a useful analytical contribution? In Chapter 1 it was seen that natural scientists and economists like to quantify and reduce, and faced with something as complex as sustainability, grasping for the same type of quantifiable indicators that work well in measuring environmental quality is a logical step for them to take. Most SIs are logical and clearly help measure important resources and processes. The lists of SIs in Chapter 1 are a useful guide for examining change, but SIs are typically a product of a western concept of sustainability based on western experience, and as Guyer points out:

> Looming behind all theories of change in Africa is the vast shadow of the comparative experience of agricultural intensification and industrial development elsewhere.

As already discussed, the problem is essentially one of vision with a suite of SIs attempting to embody it, but whose vision gets priority? This embodiment is inevitably imperfect, and it is potentially dangerous if they are reified as 'objective' measures of something called sustainability rather than for what they really are: quantitative representations of values that are partly a product of upbringing and social experiences. Sustainability cannot be measured in the same way as physical phenomena such as soil pH and organic matter. A list of apparently objective SIs may be appealing and useful, but can potentially be more constraining than liberating and the danger is that they become an end in themselves. On a more positive note, the SIs employed in this study helped provide focus. African social, economic and agricultural systems can be bewildering in their complexity so a *point de appui* is useful and lists of SIs can help provide this. However, it soon became clear that horizons needed to be expanded.

Taking this positivist attitude towards SIs, while acknowledging their limitations, the authors suggest that an examination of sustainability is best initiated at the socio-cultural level. From this portal, considerations would expand organically to agriculture, trades and all livelihoods and activities of import in the village theatre. A focus on people/community and livelihood/leisure as opposed to the village and agriculture would immediately prevent problems of restrictive spatial boundaries, and encompass access to information and the availability and ability to access options. Indeed, given that the household has become a central consideration for analysis in community studies (Guyer, 1981), it is perhaps surprising that the agricultural SI literature is still primarily focused on the more agro-ecological and geographical scales – for example the village catchment. In a paper discussing the dynamics behind diversity, Guyer makes reference to the separateness of the two sets of literature the authors consider critical in agricultural sustainability and share her insight and concern.

This means suspending confidence that the conventional measures are adequate and that new concepts such as 'diversity' can be adopted without a thorough and concentrated critique, and reconnecting literature that have not so far been considered highly relevant to one another: the social and cultural studies that explore the dynamics of human diversity within and across communities over time, and those that focus on crop ecology.

Guyer (1996)

The dearth of reference to the wealth of such 'social and cultural studies' is lamentable, particularly in relation to agriculture in a development context. As Wall (1998) wryly pointed out:

Often the sociological input is considered desirable only for its instrumental value in fulfilling the formal requirements for funding rather than for its intrinsic contributions to the essential research problem.

It is noteworthy that the term sustainability, let alone quantitative SIs, hardly features in such discussions even though the themes typically revolve around change, resilience and diversity – the very factors which underpin sustainability. The household may not always be a unit for analysis in such studies, but the centrality of people and communities is apparent. A divide is perhaps inevitable, even in this enlightened age of 'multi' and 'inter' disciplinarity, and may be attributed to the nature of the disciplines involved.

The methodological problem lies with the selection of a socio-cultural unit for analysis. The household may be a convenient unit, but is problematic nonetheless. Guyer (1981) stresses the need to be careful and not apply the concept of household in a 'rigid fashion'. Although it may be the prime focus, it is not the only consideration as households are interconnected. These constitute a resource in themselves for information and networking salient for the women traders. The importance of networks is not unique to Igala for Stone (1997) found a similar emphasis among the Tiv people in Benue State and Berry (1993) provides a general analysis for Africa. The experience in Eroke suggests the prime concern of those living there is their family. A hierarchy of levels to be considered in an analysis of sustainability may be:

MACRO-ECONOMIC/POLITICAL CONSIDERATIONS

SOCIO-CULTURAL LEVEL

SOURCES OF INCOME AND SUSTENANCE

RESOURCE BASE
(including access to information and ability to access options)

The initial focus on the community, household and family is not constrained by the spatial dimension, although a spatial consideration is inevitable and it follows that the space where the community live and work will be included in the analysis. The macro-economic and political environments also have to be considered as they impinge indirectly on their lives. Import duties on vehicle spare parts is one example that would have an effect on the traders in Eroke.

The time dimension for sustainability is a further consideration. One could apply a reasoned external definition, as with the choice of ten years for this study and others, or draw on local visions. These may be short-term, but they are in line with what people in Eroke think about 'sustainability'.

While a list of SIs can be a dangerous tool, a list of topics to explore may be a useful compromise and such a list is provided in Table 7.1. These topics are by no means original as they build on the discussions set out in Chapter 1 and the 'social and cultural studies' already mentioned. The mode of presentation matches the SI lists, and this is done with the hope of forming a bridge between the two. Quantification has a place, but so too do the more qualitative assessments. How the information gleaned within these areas is integrated into some notion of sustainability cannot be prescribed for like beauty, it is in the eye of the beholder.

Sustainability – where to now?

The association of sustainable with 'good' and unsustainable with 'bad' poses an immediate difficulty. For example, in Eroke emigration is understood as good by both the emigrants and those they leave behind, though they clearly understand that in the long term this will have repercussions for the village. An outsider may see this as an unambiguous picture of unsustainability and hence bad, but things change and in this instance the change is welcomed by those most affected. Different groups can look at the same situation and arrive at diverse views, and with sustainability the phrase 'what you see is what you get' might be more appropriately inverted to 'what you get is what you want to see'.

Using the words sustainable and unsustainable almost as labels can be problematic and counter-productive. One can readily identify change and the reasons for it, but using a word as value laden as sustainable automatically imposes subjective views and considerations. One wonders whether the word sustainable, be it as a property or an approach, is proving to be a hindrance rather than a help. The idea of carefully planning for the future or 'not cheating on the kids' are commendable, but can change and the dynamics that underpin it be looked at without resorting to value laden terms? Viewing sustainability objectively as a system property is difficult and even its more vernacular use to describe approach is equally value laden. The masking of human values within sustainability has been noted by others (for example Wall, 1998), yet the drive continues to make it 'literal, system-orientated, quantitative, predictive, stochastic and diagnostic' (Hansen, 1996).

The biologist Stephen Jay Gould of Harvard University, summarised in a sentence the tactics of those who tried to encapsulate the complexity of human intelligence in a single value called 'G' which was supposed to be largely determined by genetics (and hence a function of race):

> I have the numbers, the rigor and the objectivity: you have only hopes and emotion.

Are there parallels with those attempting to quantify and encapsulate sustainability in a single quantitative measure (S)?

Key resource people in Igalaland suggested 'viability' as an alternative to sustainability, but one wonders whether that would be as emotive as sustainability. An alternative would be to abandon any attempt to objectify and quantify sustainability as a system property, and see it instead as the spirit to survive, improve and recreate our environment despite of, and because of, the ongoing changes that call forth the creativity and inventiveness in each of us. Development intervention has to take special note of those least able to survive in such circumstances of change.

Table 7.1 Some suggested areas to explore within sustainability

Level	Suggested areas to explore (including change over time)
'Macro'	location of village and infrastructure (presence of good roads, markets etc.) linkages with others (form and function) in the village and beyond national economic and political forces growing seasons and environmental hazards (e.g. risk of flooding)
Socio-cultural	composition (gender and age profiles; relationships) change in household composition (immigration, emigration etc.) and forces that drive this change inheritance and other lineage issues leisure and management of time household income and expenditure (budgets) work activities (includes household maintenance)
Sources of income and sustenance	sources of income and expenditure agriculture and tree crops (yields, areas, prices, costs etc.) off-farm activities consumption (food and others) credit and savings associations employment opportunities
Resource base	opportunities and ability to make use of them access to information and networking (at all levels, not just within the household or village) soils and landscape labour availability (includes group labour) skills and experience plant and animal biodiversity (includes crops and their varieties) water (availability and quality) sources of fuel (e.g. firewood)

The headings are a suggested list of topics to be explored in any given situation. The precise means of exploring the 'suggested areas' will vary depending upon particular circumstances. The same sort of information (but in a different context) may be generated at each level of the analysis.

APPENDIX A

The results of Chi-square test for association (degrees of freedom are provided in parentheses) between the perceived trends of various resources and coping strategies between the late 1980s and late 1990s. The resources analysed are:

Table
8.1   Field crops (production, quality, consumption and pest/disease incidence)
8.2   Tree crops (production and quality)
8.3   Animals (ownership, disease incidence and consumption)
8.4   Firewood and water resources (availability, consumption and quality)
8.5   Coping strategies for dealing with soil fertility

The trends identified are: increase (or improved), remained the same and decrease (or worsened). The figures refer to the number of respondents who expressed that opinion about trend (the observed counts), and the figures in parentheses refer to the expected counts (based on no association). The Chi-square is testing for differences in perceived trends within the categories.

Table 8.1 Field crops

(a) crop area

Crop	Increase	Same	Decrease	Sig. (df)
groundnut	28 (-5.0)	11 (-0.08)	15 (5.08)	* (6)
cowpea	37 (-3.33)	19 (5.46)	10 (-2.13)	
bambara nut	38 (-0.5)	15 (2.08)	10 (-1.58)	
yam bean	40 (8.83)	3 (-7.46)	8 (-1.37)	
maize	23 (-0.81)	10 (4.65)	19 (-3.84)	* (2)
guinea corn/millet	26 (0.81)	1 (-4.65)	28 (3.84)	
benniseed	47 (4.29)	10 (2.88)	9 (-7.18)	** (2)
melon	19 (-4.29)	1 (-2.88)	16 (7.18)	
yam	152 (-1.69)	16 (0.27)	10 (1.42)	ns (2)
cassava	63 (1.69)	6 (-0.27)	2 (-1.42)	

(b) crop yield

Crop	Increase	Same	Decrease	Sig. (df)
groundnut	16 (-2.98)	not included	35 (2.98)	*** (3)
cowpea	14 (-9.07)	not included	48 (9.07)	
bambara nut	22 (0.05)		37 (-0.05)	
yam bean	28 (12)		15 (-12)	
maize	8 (-0.91)	not included	41 (0.91)	ns (1)
guinea corn/millet	10 (0.91)	not included	40 (-0.91)	
benniseed	36 (8.04)	not included	25 (-8.04)	** (1)
melon	8 (-8.04)	not included	27 (8.04)	
yam	77 (-13.35)	16 (1.77)	82 (11.57)	*** (2)
cassava	50 (13.35)	4 (-1.77)	17 (-11.57)	

(c) production (function of area and yield)

Crop	Increase	Same	Decrease	Sig. (df)
groundnut	21 (-2.2)	2 (2.32)	30 (4.52)	** (6)
cowpea	21 (-8.33)	5 (-0.46)	41 (8.79)	
bambara nut	28 (0.42)	5 (-0.14)	30 (-0.28)	
yam bean	32 (10.11)	7 (2.92)	11 (-13.03)	
maize	13 (-1.72)	not	34 (1.72)	ns (1)
guinea corn/millet	18 (1.72)	included	34 (-1.72)	
benniseed	40 (9.19)	not	21 (-9.19)	*** (1)
melon	9 (-9.19)	included	27 (9.19)	
yam	88 (14.44)	29 (4.81)	58 (9.63)	*** (2)
cassava	56 (-14.44)	5 (-4.81)	10 (-9.63)	

(d) quality of produce

Crop	Increase	Same	Decrease	Sig. (df)
groundnut	7 (-1.87)	24 (-1.02)	22 (2.89)	* (6)
cowpea	7 (-4.21)	30 (-1.63)	30 (5.85)	
bambara nut	9 (-1.55)	31 (1.26)	23 (0.29)	
yam bean	16 (7.63)	25 (1.39)	9 (-9.03)	
maize	not	20 (-4.25)	29 (4.25)	ns (1)
guinea corn/millet	included	29 (4.25)	21 (-4.25)	
benniseed	19 (4.12)	33 (3.24)	14 (-7.35)	** (2)
melon	4 (-4.12)	13 (-3.24)	19 (7.35)	
yam	52 (-9.18)	57 (-1.33)	66 (10.51)	** (2)
cassava	34 (9.18)	25 (1.33)	12 (-10.51)	

(e) consumption

Crop	Increase	Same	Decrease	Sig. (df)
groundnut	4 (-1.94)	not	12 (1.94)	ns (3)
cowpea	6 (0.81)	included	8 (-0.81)	
bambara nut	5 (-0.94)		11 (0.94)	
yam bean	8 (2.06)		8 (-2.06)	
maize	6 (-1.0)	8 (2.25)	6 (-1.25)	ns (6)
guinea corn	8 (1.0)	5 (-0.75)	7 (-0.25)	
millet	6 (-1.0)	5 (-0.75)	9 (1.75)	
rice	8 (1.0)	5 (-0.75)	7 (-0.25)	
benniseed	5 (0)	10 (1.5)	5 (-1.5)	ns (2)
melon	5 (0)	7 (-1.5)	8 (1.5)	
yam	3 (-5.48)	not	17 (5.48)	*** (1)
cassava	11 (5.48)	included	2 (-5.48)	

(e) crop pests and diseases

Crop	Increase	Same	Decrease	Sig. (df)
groundnut	34 (-2.75)	15 (2.75)	not	ns (2)
cowpea	55 (4.75)	12 (-4.75)	included	
maize	37 (-2)	15 (2)		
yam	40 (10.65)	39 (6.26)	13 (-16.91)	*** (2)
cassava	12 (-10.65)	19 (-6.26)	40 (16.91)	

Table 8.2 Tree crops

Tree crop	Increase	Same	Decrease	Sig. (df)
(a) number of stands				
citrus	19 (-16.08)	7 (0.41)	22 (15.67)	*** (4)
oil palm	40 (9.31)	2 (-3.77)	0 (-5.54)	
locust bean	74 (6.77)	16 (3.36)	2 (-10.13)	
(b) yield/stand				
citrus	22 (-6.8)	2 (-1.34)	21 (8.14)	** (4)
oil palm	35 (8.76)	2 (-1.05)	4 (-7.71)	
locust bean	55 (-1.96)	9 (2.39)	25 (-0.43)	
(c) production (function of number and yield)				
citrus	20 (-9.87)	not	23 (9.87)	*** (2)
oil palm	38 (10.22)	included	2 (-10.22)	
locust bean	58 (-0.35)		26 (0.35)	
(d) quality of produce				
citrus	12 (-2.4)	19 (-3.37)	14 (5.77)	*** (4)
oil palm	25 (11.88)	16 (-4.38)	0 (-7.5)	
locust bean	19 (9.48)	52 (7.75)	18 (1.73)	

Table 8.3 Change in animal ownership, disease incidence and consumption

Animal/product	Increase	Same	Decrease	Sig. (df)
(a) ownership (number/household)				
goats	21 (1.49)	not	39 (-1.49)	ns (1)
hens	19 (-1.49)	included	44 (1.49)	
(b) disease incidence				
goats	35 (-2.29)	17 (0.48)	7 (1.81)	ns (2)
hens	44 (2.29)	18 (-0.48)	4 (-1.81)	
(c) consumption				
guinea fowl/bush meat	2 (-5.73)	14 (-1.46)	14 (7.2)	*** (4)
sheep/goat	4 (-3.47)	20 (5.05)	5 (-1.58)	
chicken/eggs	19 (9.21)	16 (-3.59)	3 (-5.62)	

Table 8.4 Firewood and water resources

(1) change in availability, consumption and quality

Resource	Increase	Same	Decrease	Sig. (df)
(a) firewood availability consumption	2 (-27.5) 57 (27.5)	not included	65 (27.5) 10 (-27.5)	*** (1)
(b) water availability consumption quality	64 (0.33) 65 (1.33) 62 (-1.67)	not included	7 (-0.33) 6 (-1.33) 9 (1.67)	ns (2)

(2) availability and distance of collection of important tree species for firewood from the late 1980s to 1998.

Results based on the views of 20 female respondents. The species have been divided into 2 groups:

1. Species which are planted and maintained by the people (9 species in total e.g. mango, cashew, locust bean, guava and oil palm).

2. Species which are not planted i.e. they grow wild (total of 38 species).

Group of tree species	Change in species availability			Change in distance of collection	
	Increase	Same	Decrease	Same	Longer
planted	135 (100.49)	26 (-14.86)	7 (-85.63)	160 (87.63)	9 (-87.63)
not planted	39 (-100.49)	180 (14.86)	460 (85.63)	204 (-87.63)	477 (87.63)
Sig. (df)		*** (2)		*** (1)	

Table 8.4 continued

(3) Selection of best and worst trees for firewood, and how their availability and distance of collection have changed between the late 1970s and 1998.

In this table the four 'best' and four 'worst' species have been presented based on the views of 10 female respondents. Change in availability is based on the views of 20 female respondents, but not all expressed an opinion for each species.

Species	Preferences		Change in availability		
	Best	Worst	Increase	Same	Decrease
(a) Best firewood species					
'Enache'	9	0	0	9	10
'Ode'	8	0	0	9	10
'Aya'	6	0	0	7	12
'Ayigele'	5	0	0	2	17
Totals (best species)			0 (-9.06)	27 (-4.71)	49 (13.77)
(b) Worst firewood species					
'Odologwu'	0	9	7	10	2
'Ugba'	0	8	8	10	1
'Agba'	0	4	0	5	14
'Ejiji'	0	3	3	11	4
Totals (worst species)			18 (9.06)	36 (4.71)	21 (-13.77)
Sig. (df)				*** (2)	

Table 8.5 Coping strategies for dealing with problems of soil fertility

Strategy	Increase	Same	Decrease	Sig. (df)
fallow periods	30 (-7.26)	13 (-2.38)	26 (9.64)	*** (10)
fertilizer use	7 (-27.56)	1 (-13.26)	56 (40.82)	
compost/manure	39 (17.94)	0 (-8.69)	0 (-9.25)	
burying of crop residue	6 (-30.26)	62 (46.62)	1 (-15.36)	
planting legume trees	37 (15.94)	2 (-6.69)	0 (-9.25)	
planting legume crops	70 (32.2)	0 (-15.6)	0 (-16.6)	

Results taken from the 1997/98 production surveys (40 male and 33 female respondents).

APPENDIX B

Calculation of the consumption of foodstuff and land necessary to sustain one individual household (HH) member per annum based on a linear regression analysis. The calculation involves two steps:

Step 1. An estimation of consumption/household member based on survey data of consumption/household/annum.

Foodstuff	Consumption/ Individual HH member (kg/annum)	SD of regression	Regression with HH size Sig. and error df in parentheses	
			Intercept calculated	Intercept = 0
rice	14.1	1.58	* (17)	*** (18)
cowpea	45.7	6.4	*** (18)	*** (19)
bambara nut	4.6	2.28	ns (10)	ns (11)
pigeon pea	14.3	2.81	* (9)	*** (10)
benniseed	3.1	0.46	ns (18)	*** (19)
melon	2.1	0.44	ns (17)	*** (18)
maize grain	41.3	5.09	* (18)	*** (19)
cassava tubers	240.5	34.52	ns (18)	*** (19)
yam tubers	53.1	7.38	ns (18)	*** (19)

HH size was the independent variable and HH foodstuff consumption the dependent variable. Two regression models were fitted. The first allows for an intercept to be calculated, while the second assumes that the intercept was zero (i.e. consumption was zero for a HH size of zero). The annual consumption of foodstuff/HH member is based on the regression slope assuming that the intercept was zero (SD = standard deviation of the slope). In each case * $P < 0.05$, ** $P < 0.01$ and *** $P < 0.001$.

Step 2. An estimation of land area required to sustain one household member/year based on plot yields and the calculation in Step 1.

Upper and lower estimates are based on means and 95% confidence intervals for both consumption/HH member/annum and crop yield.

Crop (variety)	Lower land estimate (ha)	Upper land estimate (ha)
rice	0.013	0.046
cowpea	0.101	0.252
bambara nut	0	0.055
benniseed	0.018	0.241
maize (local)	0.038	0.112
maize (NCV)	0.028	0.07
cassava (local)	0.017	0.067
cassava (NCV)	0.014	0.046

APPENDIX C

Results of Mann-Whitney tests applied to rank data for occupation, income and expenditure. N refers to the number of respondents mentioning that category, and the Mann-Whitney test was applied to neighboring categories. The data were collected in a 1997 survey with 61 respondents (25 male and 36 female). The median rank (1 is high) is presented at the foot of each column of tests.

ns = neighboring categories not significant at 5%; * $P < 0.05$; ** $P < 0.01$; *** $P < 0.001$

(a) Male occupation.

Source	N	Farming	Trading	Other	Hunting	Palm tapping
Farming	25		*	**	***	***
Trading	13			ns	ns	ns
Other	9				ns	ns
Hunting	10					ns
Palm tapping	9					
Median		1	2	2	2.5	3

(b) Female occupation.

Source	N	Trading	Other	Farming	Soap making	Brewing
Trading	45		*	***	**	***
Other	33			ns	ns	*
Farming	41				ns	**
Soap making	8					ns
Brewing	7					
Median		1	2	2	3	3

Appendix C 233

(c) Sources of male income.

Source	N	Trad.	Farm.	Other	Tree crops	Remit.	Animals
Trading	14		ns	ns	*	***	***
Farming	25			ns	*	***	***
Other	11				*	***	***
Tree crops	23					***	***
Remittances	11						ns
Animals	17						
Median		1.5	2	2	3	4	4

(d) sources of female income.

Source	N	Trad.	Other	Farm.	Tree crops	Remit.	Animals
Trading	45		ns	***	***	***	***
Other	34			***	***	***	***
Farming	40				***	***	***
Tree crops	42					ns	**
Remittances	28						ns
Animals	41						
Median		1	1	3	3	4	4

(e) male expenditure.

	N	Food	Hth	Sch	Frm	Cth	Funer	Tran	Wat
Food	23		ns	ns	***	***	***	***	***
Heal	25			ns	**	***	***	***	***
Sch	23				*	**	***	***	***
Farm	24					ns	**	***	***
Cloth	25						ns	***	***
Funer	22							*	***
Trans	21								**
Wat	15								
Median		2	2	3	4	5	5	7	8

(f) female expenditure.

	N	Hth	Food	Sch	Cth	Funer	Frm	Tran	Wat
Heal	48		ns	ns	*	***	***	***	***
Food	45			ns	ns	***	***	***	***
Sch	36				ns	***	***	***	***
Cloth	50					***	***	***	***
Funer	36						ns	***	***
Farm	36							**	***
Trans	37								ns
Wat	40								
Median		2	2	3	3	4.5	5	6	7

Hth	health costs
Sch	school fees
Cth	clothing
Funer	funeral expenses
Frm	farming
Tran	transport
Wat	water

(g) ranking of male income compared with ranking of female income.

		Median rank	
Source	Sig.	Male	Female
Farming	**	2	3
Trading	ns	1.5	1
Remittances	*	4	4
Tree crop	**	3	3
Animals	ns	4	4
Other	ns	2	1

(h) ranking of male expenditure compared with ranking of female expenditure.

		Median rank	
Item	Sig.	Male	Female
Food	ns	2	2
Health	ns	2	2
School fees	ns	3	3
Farming	*	4	5
Clothing	**	5	3
Funerals	ns	5	4.5
Transport	ns	7	6
Water	**	8	7

References

Abasiekong, E. M. (1981), 'Pledging oil palms: A case study on obtaining rural credit in Nigeria', *African Studies Review,* vol. 24 (1), pp. 73-82.

Adegboye, R. O. (1983), 'Procuring loans by pledging cocoa trees' in J. A. Von Pische, D. W. Adams and G. Donald (eds.), *Rural finance markets in developing countries: Their uses and abuses,* John Hopkins University Press, Baltimore.

Ainsworth, E. (1989), 'LISA men have called you', *Farm Journal,* vol. 113 (February), pp. 1.

Allen, P., Van Dusen, D., Lundy, L. and Gliessman, S. (1991), 'Integrating social, environmental and economic issues in sustainable agriculture', *American Journal of Alternative Agriculture,* vol. 6, pp. 34-39.

Avery, D. T. (1995), *Saving the planet with pesticides and plastics: the environmental triumph of high-yield farming,* Hudson Institute, Indianapolis.

Beets, W. C. (1990), *Raising and sustaining productivity of smallholder farming systems in the tropics: a handbook of sustainable agricultural development,* AgBe, Alkmaar, Holland.

Begon, M. (1990), *Ecological food production: a food production policy for Britain,* Institute for Public Policy Research, London.

Bell, S. and Morse, S. (1998), *Sustainability indicators. Measuring the immeasurable,* Earthscan, London.

Berry, S. (1993), *No condition is permanent. The social dynamics of agrarian change in Sub-Saharan Africa,* University of Wisconsin Press, Madison.

Blaikie, P. (1995), 'Understanding environmental issues', in S. Morse and M. Stocking (eds.), *People and environment,* UCL Press, London, pp. 1-30.

Boserup, E. (1965), *The conditions of agricultural growth; the economics of agrarian change under population pressure,* Allen and Unwim, London.

Brader, L. (1988), 'Needs and directions for plant protection in developing countries: The FAO view', *FAO Plant Protection Bulletin,* vol. 36 (1), pp. 3-8.

Briggs, D. and Courtney, F. (1989), *Agriculture and environment. The physical geography of temperate agricultural systems*, Longman, Harlow, UK.
Carson, R. (1962), *Silent spring*, Penguin, London.
Casley, D. J. and Lury, D. A. (1982), *Monitoring and evaluation of agriculture and rural development projects*, John Hopkins University Press for the World Bank, Baltimore.
Chambers, R. (1991), *Rural development: Putting the last first*, Intermediate Technology, London.
Clarke, J. (1981), 'Households and the political economy of small-scale cash crop production in South-Western Nigeria', *Africa,* vol. 51 (4), pp. 807-823.
Clayton, M. H. and Radcliffe, N. J. (1996), *Sustainability. A systems approach*, Earthscan, London.
Cleave, J. H. (1974), *African farmers: Labour use in the development of smallholder agriculture*, Praeger, New York.
Common, M. and Perrings, C. (1992), 'Towards an ecological economics of sustainability', *Ecological Economics,* vol. 6, pp. 7-34.
Conway, G. R. (1986), *Agroecosystem analysis for research and development*, Winrock International, Bangkok.
Cross, N. and Barker, R. (1953), *At the deserts edge. Oral histories from the Sahel*, Panos, London.
Csapo, M. (1983), 'Universal Primary Education in Nigeria: It's problems and implications', *African Studies Review*, vol. 26 (1), pp. 91-106.
Dennis, C. (1983), 'Capitalist development and women's work: a Nigerian case study', *Review of African Political Economy*, vol. 27/28, pp. 109-119.
Dobbs, T. L., Becker, D. L and Taylor, D. C. (1991), 'Sustainable agriculture policy analysis: South Dakota on-farm case studies', *Farming Systems Research and Extension*, vol. 2, pp. 109-124.
Dunlap, R. E., Beus, C. E., Howell, R. E. and Waud, J. (1992), 'What is sustainable agriculture? An empirical examination of faculty and farmer definitions', *Journal of Sustainable Agriculture*, vol. 3 (1), pp. 5-39.
Ehui, S. K. and Spencer, D. S. C. (1993), 'Measuring the sustainability and economic viability of tropical farming systems: a model from sub-Saharan Africa', *Agricultural Economics*, vol. 9, pp. 279-296.

Ezumah, N. N. and Di Domenico, C. M. (1995), 'Enhancing the role of women in crop production: A case study of Igbo women in Nigeria', *World Development*, vol. 23 (10), pp. 1731-1744.

Flinn, J. C. and Zuckerman, P. S. (1981), 'Production, income and expenditure patterns of Yoruba smallholders', *Africa*, vol. 51 (4), pp. 825-835.

Flora, C. B. (1992), 'Building sustainable agriculture: A new application of Farming Systems Research and Extension', *Journal of Sustainable Agriculture*, vol. 2 (3), pp. 37-49.

Francis, C. A. (1989), 'Biological efficiencies in multiple-cropping systems', *Advances in Agronomy*, vol. 42, pp. 1-42.

Frans, R. (1993), 'Sustainability of high-input cropping systems: the role of IPM', *FAO Plant Protection Bulletin*, vol. 41 (3/4), pp. 161-170.

Fresco, L. O. and Kroonenberg, S. B. (1992), 'Time and spatial scales in ecological sustainability', *Land Use Policy*, vol. 9, pp. 155-168.

Goldman, A. (1995), 'Threats to sustainability in African agriculture: Searching for appropriate paradigms', *Human Ecology*, vol. 23 (3), pp. 291-334.

Goldman, A. (1996), 'Pest and disease hazards and sustainability in African agriculture', *Experimental Agriculture*, vol. 32, pp. 199-211.

Goldman, A. and Smith, J. (1995), 'Agricultural transformations in India and Northern Nigeria: Exploring the nature of Green Revolutions', *World Development*, vol. 23 (2), pp. 243-263.

Gould, S. J. (1990), *An urchin in the storm*, Penguin, London.

Grieg-Smith, P., Frampton, G. and Hardy, T. (1992), *Pesticides, cereal farming and the environment*, HMSO, London.

Guyer, J. I. (1981), 'Household and community in African studies', *African Studies Review*, vol. 24 (2/3), pp. 87-137.

Guyer, J. I. (1992), 'Small change: Individual farm work and collective life in a western Nigerian Savanna town, 1969-88', *Africa*, vol. 62 (4), pp. 465-488.

Guyer, J. I. (1996), 'Diversity at different levels: Farm and community in West Africa', *Africa*, vol. 66 (1), pp. 71-89.

Guyer, J. I. (1997), *An African niche economy*, Edinburgh University Press, Edinburgh.

Hansen, J. W. (1996), 'Is agricultural sustainability a useful concept?', *Agricultural Systems*, vol. 50, pp. 117-143.

Harrington, L. (1992a), 'Measuring sustainability: issues and alternatives', in W. Hiemstra, C. Reijntjes and E. van der Werf (eds.), *Let farmers judge: Experiences in assessing the sustainability of agriculture*, Intermediate Technology, London, pp. 3-16.

Harrington, L. (1992b), 'Measuring sustainability: issues and alternatives', *Journal for Farming Systems Research-Extension*, vol. 3 (1), pp. 1-20.

Hendrix, P. F., Coleman, D. C. and Crossley, D. A. Jr. (1992), 'Using knowledge of soil nutrient cycling processes to design sustainable agriculture', *Journal of Sustainable Agriculture*, vol. 2 (3), pp. 63-82.

Innis, D. Q. (1997), *Intercropping and the scientific basis of traditional agriculture*, Intermediate Technology Publications, London.

International Union for the Conservation of Nature (1991), *Caring for the Earth: a strategy for sustainable living*, IUNC, Gland, Switzerland.

Izac, A-M., N. and Swift, M. J. (1994), 'On agricultural sustainability and its measurement in small-scale farming in sub-Saharan Africa', *Ecological Economics*, vol. 11, pp. 105-125.

Jansen, D. M., Stoorvogel, J. J. and Schipper, R. A. (1995), 'Using sustainability indicators in agricultural land use analysis: an example from Costa Rica', *Netherlands Journal of Agricultural Science*, vol. 43, pp. 61-82.

Jodha, N. S. (1989), 'Mountain agriculture: Search for sustainability', paper presented at the International Symposium on *Impacts of Farming Systems Research and Extension on Sustainable Agriculture,* University of Arkansas, Fayetteville.

Kidd, C. V. (1992), 'The evolution of sustainability', *Journal of Agricultural and Environmental Ethics*, vol. 5 (1), pp. 1-26.

Kremer, A. (1994), 'Strategic issues in West African IPM.', in *IPM Implementation Workshop for West Africa*, Natural Resources Institute, Chatham, UK, pp. 105-110.

Kumar, K. (1993), 'An overview of rapid rural appraisal methods in development settings', in K. Kumar (ed.), *Rapid appraisal methods*, World Bank, Washington DC, pp. 8-22.

Lawuyi, T. and Falola, T. (1992), 'The instability of the Naira and social payment among the Yoruba', *Journal of Asian and African Studies*, vol. 27 (3-4), pp. 216-228.

Lehman, H., Clark, E. A. and Weise, S. F. (1993), 'Clarifying the definition of sustainable agriculture', *Journal of Agricultural and Environmental Ethics*, vol. 6 (2), pp. 127-143.

Lele, S. M. (1991), 'Sustainable development: a critical review', *World Development*, vol. 19 (6), pp. 607-621.

Lynam, J. K. and Herdt, R. W. (1989), 'Sense and sensibility: sustainability as an objective in international agricultural research', *Agricultural Economics*, vol. 3, pp. 381-398.

MacRae, R. J., Hill, S. B., Henning, J. and Mehuys, G. R. (1989), 'Agricultural science and sustainable agriculture: a review of the existing scientific barriers to sustainable food production and potential solutions', *Biological Agriculture and Horticulture*, vol. 6, pp. 173-219.

Manyong, V. M. and Degand, J. (1997), 'Measurement of the sustainability of African smallholder farming systems: Case study of a systems approach', *IITA Research*, vol. 14/15, pp. 1-6.

McNamara, N. and Morse, S. (1996), *Developing on-farm research*, On Stream, Cork.

McNamara, N. and Morse, S. (1998*)*, *Developing financial services. A case against sustainability*, On Stream, Cork

Miller, F. P. and Wali, M. K. (1995), 'Soils, land use and sustainable agriculture: a review', *Canadian Journal of Soil Science*, vol. 75, pp. 413-422.

Morse, S. and Buhler, W. (1997), *Integrated pest management. Ideals and realities in developing countries*, Lynne Reinner, Boulder, Colarodo.

Neher, D. (1992), 'Ecological sustainability in agricultural systems: Definition and measurement', *Journal of Sustainable Agriculture*, vol. 2 (3), pp. 51-61.

Netting, R. McC. (1993), *Smallholders, householders. Farm families and the ecology of intensive, sustainable agriculture*, Stanford University Press, Stanford.

Netting, R. McC., Stone, M. P. and Stone, G. D. (1989), 'Kofyar cash-cropping: Choice and change in indigenous agricultural development', *Human Ecology*, vol. 17 (3), pp. 299-319.

Netting, R. McC. and Stone, M. P. (1996), 'Agro-diversity on a farming frontier: Kofyar smallholders on the Benue plains of central Nigeria', *Africa*, vol. 66 (1), pp. 52-70.

Ofori, F. and Stern, W. R. (1987), 'Cereal-legume intercropping systems', *Advances in Agronomy*, vol. 41, pp. 41-90

Olsen, R. K. (1992), 'The future context of sustainable agriculture: Planning for uncertainty', *Journal of Sustainable Agriculture*, vol. 2 (3), pp. 9-20.

Otzen, U. (1993), 'Reflections on the principles of sustainable agricultural development', *Environmental Conservation*, vol. 20 (4), pp. 310-316.
Pearce, D. (1993), *Blueprint 3. Measuring sustainable development*, Earthscan, London
Peil, M., Ekpenyong, S. K. and Oyeneye, O. Y. (1988), 'Going home: Migration careers of Southern Nigerians', *International Migration Review*, vol. 22 (4), pp. 563-585.
Penfold, C. M., Miyan, M. S., Reeves, T. G. and Grierson, I. T. (1995), 'Biological farming for sustainable agricultural production', *Australian Journal of Experimental Agriculture*, vol. 35, pp. 849-856.
Pimental, D., Berardi, G. and Fast, S. (1983), 'Energy efficiency of farming systems: Organic and conventional agriculture', *Agriculture, Ecosystems and Environment*, vol. 9, pp. 359-372.
Reijntjes, C., Haverkort, B. and Waters-Bayer, A. (1992), *Farming for the future. An introduction to low-external-input and sustainable agriculture*, Macmillan Press, London.
Rennings, K. and Wiggering, H. (1997), 'Steps towards indicators of sustainable development: linking economic and ecological concepts', *Ecological Economics*, vol. 20, pp. 25-36.
Richards, P. (1985), *Indigenous agricultural revolution*, Hutchinson, London.
Richards, P. (1990), 'Local strategies for coping with hunger: Central Sierra Leone and Northern Nigeria compared', *African Affairs*, vol. 89 (355), pp. 265-275.
Schaller, N. (1993), 'The concept of agricultural sustainability', *Agriculture, Ecosystems and Environment*, vol. 46, pp. 89-97.
Seibel, H. D. (1983), *A preliminary appraisal of state development projects in the Western Zone and of the Farmer Council project of Idah, Benue State*. Unpublished report.
Slesser, M. (1984), 'Energy use in the food-producing sector of the European Economic Community' in G. Stanhill (ed.), *Energy and agriculture*, Springer-Varlag, Berlin, pp. 132-153.
Smith, G. R. (1998), 'Are we leaving the community out of rural community sustainability?' *International Journal of Sustainable Development and World Ecology*, vol. 5, pp. 82-98.
Smith, C. S. and McDonald, G. T. (1998), 'Assessing the sustainability of agriculture at the planning stage', *Journal of Environmental Management*, vol. 52, pp. 15-37.

Spencer, D. S. C. and Swift, M. J. (1992), 'Sustainable agriculture: Definition and measurement', in K. Mulongoy, M. Gueye and D. S. C. Spencer (eds.), *Biological nitrogen fixation and sustainability of tropical agriculture*, John Wiley and Sons, Chichester, pp. 15-24.

Stanhill, G. (1984), 'Agricultural labour: From energy source to sink', in G. Stanhill (ed.), *Energy and agriculture*, Springer-Varlag, Berlin, pp. 113-129.

Steinbeck, J. (1939), *The Grapes of Wrath*, Heinemann, London.

Stocking, M. (1994), 'Soil erosion and conservation: A place for soil science?', in J. K. Syers and D. L. Rimmer (eds), *Soil science and sustainable land management*, CAB International, Wallingford, pp. 40-58.

Stocking, M. (1998), 'Soil and water conservation' in C. C. Webster and P. N. Wilson (eds.), *Agriculture in the tropics*, Blackwell Science, Oxford, pp. 82-112.

Stockle, C. O., Papendick, R. I, Saxton, K. E., Campbell, G. S. and van Evert, F. K. (1994), 'A framework for evaluating the sustainability of agricultural production systems', *American Journal of Alternative Agriculture*, vol. 9 (1 and 2), pp. 45-50.

Stone, G. D. (1997), 'Predatory sedentism Intimidation and intensification in the Nigerian savanna', *Human Ecology*, vol. 25 (2), pp. 223-242.

Stone, G. D. (1998), 'Keeping the home fires burning: The changed nature of householding in the Kofyar homeland', *Human Ecology*, vol. 26 (2), pp. 239-265.

Stone, G. D., Netting, R. McC. and Stone, M. P. (1990), 'Seasonality, labor scheduling, and agricultural intensification in the Nigerian savanna', *American Anthropologist*, vol. 92 (1), pp. 7-23.

Stone, M. P., Stone, G. D. and Netting, R. McC. (1995), 'The sexual division of labor in Kofyar agriculture', *American Ethnologist*, vol. 22 (1), pp. 165-186.

Swift, M. J. and Woomer, P. (1993), 'Organic matter and the sustainability of agricultural systems: Definition and measurement', in K. Mulongov and R. Merckx (eds.), *Soil organic matter dynamics and sustainability of tropical agriculture*, John Wiley and Sons, Chichester, pp. 3-17.

Swindell, K. (1984), 'Farmers, traders and labourers: Dry season migration from north-west Nigeria, 1900-33', *Africa*, vol. 54 (1), pp. 3-19.

Taylor, D. C., Mohamed, Z. A., Shamsudin, M. N., Mohayidin, M. G. and Chiew, E. F. C. (1993), 'Creating a farmer sustainability index: A Malaysian case study', *American Journal of Alternative Agriculture*, vol. 8 (4), pp. 175-184.

Tisdall, C. (1996), 'Economic indicators to assess the sustainability of conservation farming projects: an evaluation', *Agriculture, Ecosystems and the Environment*, vol. 57, pp. 117-131.

Tivy, J. (1990), *Agricultural ecology*, Longman, Harlow, UK.

Torquebiau, E. (1992), 'Are tropical agroforestry home gardens sustainable?', *Agriculture, Ecosystems and Environment*, vol. 41, pp. 189-207.

Trenbath, R. B. (1974), 'Diversify or be damned?', *The Ecologist*, vol. 5, pp. 76-83.

Wall, E. (1998), 'The company of strangers: Sociology in trans-disciplinary research', *Canadian Journal of Sociology*, vol. 23 (2-3), pp. 281-300.

World Commission on Environment and Development (1987), *Our common future*, Oxford University Press, Oxford.

Yunlong, C. and Smit, B. (1994), 'Sustainability in agriculture: a general review', *Agriculture, Ecosystems and Environment*, vol. 49, pp. 299-307.

Index

Abeokuta, 68
Abo Eroke, 49, 76, 106, 113, 126
Abo Inele, 49, 76, 106, 113, 126
Abo Ogbodo, 49, 76
Abo Ojoche, 49, 76, 106, 113
Abo Ojuwo, 49, 76, 185, 204
Abuja, 122, 193
acrisol, 19
active ingredient (ai), 21, 34
adakpo, 94, 95, 148, 149, 167, 170
Ado-Ekiti, 68
adult, 58, 77, 87, 93, 97, 101, 126, 195
Africa, 12, 22, 23, 26, 51, 82, 101, 190, 216, 217
Africa, Sub-Saharan, 26, 36
Africa, West, 49, 91, 94, 95, 101, 103, 114, 124, 133, 146
Agenda 21, 24
agri-business, 7, 22, 137
agricultural education, 35
agricultural intensification, 216
agriculture, alternative, 7
agriculture, low input (LISA), 7
agriculture, organic, 7, 8
agriculture, system, 6, 7, 10, 11, 17, 22, 28, 135, 216
agro-forestry, 27
agro-inputs, 22, 180, 199
alternative agriculture, 7
aluminium, 36, 192, 212
analysis of variance (ANOVA), 66, 156, 162, 163, 204
Ankpa, 74, 75, 76, 82, 84, 89, 90, 112, 119, 124, 125, 179, 181, 192, 193, 200
Area Traditional Council (ATC), 86, 106
Attah, 83, 86, 106
ayilo, 94, 95, 148, 167, 170

bambara nut, 64, 89, 91, 109, 135, 136, 137, 138, 143, 151, 152, 156, 157, 160, 162, 171, 172, 173, 222, 223, 224, 230, 231
banana, 92, 110, 136, 140, 164
barley, 17
benniseed, 89, 109, 120, 135, 136, 137, 138, 139, 140, 141, 151, 153, 156, 157, 160, 162, 171, 172, 173, 174, 179, 222, 223, 224, 230, 231
Biocide index, 21, 25, 32
biodiversity, 12, 22, 36, 51, 82, 137, 220
biological oxygen demand (BOD), 38
biomass, 36, 37
blacksmith, 60, 119, 120
Borneo, 17, 18
Burundi, 30, 41
bush mango, 92, 110, 151
bush meat, 152, 161, 226

cabbage, 29, 40
Calabar, 74, 117, 118, 124, 133
case study, 26, 29, 30, 47, 49, 187
cashew, 92, 110, 136, 140, 150, 164, 227

cassava, 61, 64, 89, 90, 91, 92, 94, 99, 109, 116, 120, 129, 135, 136, 137, 138, 139, 140, 141, 143, 144, 145, 146, 147, 151, 152, 153, 154, 156, 157, 158, 159, 160, 163, 171, 172, 173, 174, 181, 186, 201, 215, 222, 223, 224, 230, 231
Catholic Diocese of Idah, 81
change, 3, 4, 5, 7, 16, 19, 24, 29, 32, 47, 48, 51, 52, 55, 57, 58, 61, 62, 63, 64, 66, 70, 72, 77, 86, 97, 98, 101, 105, 135, 138, 139, 140, 141, 144, 152, 153, 154, 155, 160, 166, 171, 183, 185, 189, 193, 194, 196, 205, 209, 211, 214, 216, 217, 218, 219, 220, 227
Chief, *Attah*, 83, 86, 106
Chief, g*ago*, 85, 86, 106, 184
Chief, m*adaki*, 84, 85, 86, 97, 101, 106, 113, 115, 126, 184
child, 102, 105, 122, 126, 127, 192
Chi-square, 66, 138, 197, 221
citrus, 92, 110, 122, 140, 160, 164, 225
clan, 51, 83, 84, 85, 93, 97, 106, 116, 120, 123, 125, 198
clinic, 180, 195, 202
cocoa, 119, 137, 151, 154, 192, 193
coconut, 136
cocoyam, 92, 109, 136, 151
coffee, 122, 137, 154, 192
community studies, 23, 49, 52, 216
compost, 2, 18, 99, 143, 166, 229
compound, 51, 58, 60, 63, 64, 82, 94, 99, 101, 116, 119, 121, 122, 123, 127, 129, 142, 143, 190
coping strategies, 52, 71, 143, 154, 166, 214, 221
corruption, 80, 182, 206
Council, Area Traditional (ATC), 86, 106

Council, Local Government Chieftancy, 85
cowpea, 64, 90, 91, 92, 109, 116, 120, 128, 135, 136, 137, 138, 143, 144, 145, 151, 152, 153, 155, 156, 157, 160, 162, 171, 172, 173, 174, 186, 222, 223, 224, 230, 231
credit, 38, 51, 55, 82, 95, 128, 183, 184, 187, 188, 189, 194, 195, 204, 206, 213, 220
crop mixtures, 157
crop protection, 20, 29, 136, 145, 155, 174
crop residue, 18, 35, 166, 229
crop rotation, 10, 40
crop varieties, local, 97, 136, 140, 173
crop varieties, new (NCV), 56, 82, 156, 231
cropping, 12, 14, 15, 28, 43, 136, 144
cropping, cycle, 14
cropping, system, 6, 12, 13, 14, 22, 23, 26, 27, 30, 43, 134, 135, 137, 143
crops, bambara nut, 64, 89, 91, 109, 135, 136, 137, 138, 143, 151, 152, 156, 157, 160, 162, 171, 172, 173, 222, 223, 224, 230, 231
crops, barley, 17
crops, benniseed, 89, 109, 120, 135, 136, 137, 138, 139, 140, 141, 151, 153, 156, 157, 160, 162, 171, 172, 173, 174, 179, 222, 223, 224, 230, 231
crops, cabbage, 29, 40
crops, cassava, 61, 64, 89, 90, 91, 92, 94, 99, 109, 116, 120, 129, 135, 136, 137, 138, 139, 140, 141, 143, 144, 145, 146, 147, 151, 152, 153, 154, 156, 157, 158, 159, 160, 163, 171, 172, 173, 174, 181, 186, 201, 215, 222, 223, 224, 230, 231

crops, cocoyam, 92, 109, 136, 151
crops, cowpea, 64, 90, 91, 92, 109, 116, 120, 128, 135, 136, 137, 138, 143, 144, 145, 151, 152, 153, 155, 156, 157, 160, 162, 171, 172, 173, 174, 186, 222, 223, 224, 230, 231
crops, groundnut, 91, 109, 135, 137, 143, 144, 152, 160, 186, 222, 223, 224
crops, guinea corn, 89, 90, 91, 95, 128, 135, 137, 139, 157, 160, 222, 223, 224
crops, millet, 89, 90, 91, 95, 109, 120, 135, 136, 137, 139, 160, 222, 223, 224
crops, pigeon pea, 91, 109, 151, 230
crops, rice, 17, 42, 61, 91, 92, 109, 151, 153, 156, 171, 172, 186, 201, 224, 230, 231
crops, soybean, 152, 153, 186
crops, wheat, 12, 17
crops, yam, 61, 91, 92, 96, 109, 119, 120, 136, 137, 138, 139, 140, 141, 142, 143, 144, 147, 152, 153, 157, 160, 169, 171, 174, 201, 214, 222, 223, 224, 230
crops, yam bean, 109, 137, 138, 140, 141, 142, 143, 157, 160, 222, 223, 224
cultivation intensity, R-factor, 19, 141, 165

DDT, 19
Diocesan Development Services (DDS), 2, 3, 4, 47, 48, 49, 56, 57, 61, 69, 73, 81, 82, 92, 93, 121, 144, 152, 155, 179, 180, 183, 184, 185, 186, 187, 188, 189, 195, 202, 206, 208, 212, 213, 214
diseases, 13, 36, 62, 71, 90, 143, 144, 150, 153, 160, 209, 214, 224
diversity, 4, 8, 9, 12, 20, 23, 31, 37, 52, 77, 130, 136, 137, 138, 140, 196, 216, 217
division of labour, 145

Earth Summit, Rio, 5, 24
ecological food production, 7
ecology, 6, 217
economic viability, 8, 12, 14, 15, 16
education, primary, 64, 82, 88, 117, 119, 120, 121, 122, 123, 124, 125, 126, 127, 132, 139, 180, 196
education, secondary, 64, 116, 117, 122, 123, 124, 126, 127, 132, 147
education, tertiary, 99, 132
egusi melon, 109, 137, 138, 139, 140, 141, 179
elder, 58, 121, 125
emigration, 88, 131, 193, 211, 218, 220
energy, flow, 23, 32
energy, input, 17, 35
energy, output, 17
energy, ratio, 17, 39
Enugu, 76, 117
environment, 5, 7, 8, 10, 16, 17, 19, 21, 22, 25, 26, 27, 65, 114, 143, 219
Equatorial Guinea, 117, 118, 121, 133
equitability, 9
erosion, 10, 18, 19, 28, 37, 38, 40, 97, 134, 192, 212
eutrophication, 38
expenditure, 60, 61, 64, 68, 70, 71, 72, 77, 180, 207, 220, 232, 234, 235
externalities, 16

fallow, 19, 83, 87, 135, 142, 143, 166, 174, 215, 229
family, 8, 50, 56, 58, 61, 88, 89, 90, 93, 98, 99, 100, 101, 104, 116, 117, 120, 122, 123, 124, 125, 129, 142, 151, 154, 178, 182, 184, 194, 208, 213, 217, 218
family, extended, 50, 120, 130

Farmer Council (FC), 57, 82, 107, 183, 184, 185, 186, 187, 188, 206, 208
Farmer Sustainability Index (FSI), 29
ferraisol, 19
fertilizer, 2, 5, 8, 11, 13, 17, 18, 19, 24, 34, 35, 40, 137, 139, 143, 144, 166, 229
financial institutions, *esusu*, 103
financial institutions, *oja*, 103, 104, 129, 181, 182, 183, 188, 190, 195, 206
firewood, 51, 62, 63, 71, 110, 111, 149, 150, 154, 155, 171, 174, 220, 227, 228
fish, 90, 128, 151
focus group, 54, 55, 61, 65
forestry, 7
fossil fuels, 7, 17
fuel wood, 37, 96, 105
Fulani, 83
futurity, 9, 25

gago, 85, 86, 106, 184
gender issues, 51
genetics, 219
glucides, 41
gmelina, 93, 150
Gross Domestic Product (GDP), 34
groundnut (peanut), 91, 109, 135, 137, 143, 144, 152, 160, 186, 222, 223, 224
guava, 92, 110, 136, 140, 150, 164, 227
guinea corn (sorghum), 89, 90, 91, 95, 128, 135, 137, 139, 157, 160, 222, 223, 224
guinea fowl, 152, 161, 226

Hausa, 203

health, 39, 40, 51, 59, 61, 64, 70, 71, 73, 78, 79, 80, 82, 87, 88, 95, 102, 104, 116, 119, 142, 150, 180, 194, 195, 200, 206, 234
health care, 59, 61, 64, 70, 82, 87, 88, 104, 180, 200, 206
herbalist, 60, 115, 124, 126, 127, 128
hospital, 100, 104, 117, 118, 180, 202
household, 36, 37, 49, 50, 51, 56, 57, 58, 59, 60, 61, 62, 63, 64, 70, 71, 72, 73, 86, 87, 88, 89, 91, 92, 94, 98, 99, 100, 101, 102, 103, 107, 108, 116, 119, 120, 121, 122, 127, 130, 131, 132, 134, 135, 143, 147, 148, 149, 151, 152, 153, 154, 155, 167, 171, 178, 179, 190, 194, 199, 205, 206, 215, 216, 217, 218, 220, 226, 230, 231
household, head of, 50, 58, 59, 61, 72, 100, 121, 127, 132, 149
human intelligence, 219
human resource base, 51
hunger, 68, 154, 174
hunting, 83, 84, 88, 89, 90, 192

Ibadan, 68, 74, 90, 193
Idah, 2, 74, 81, 83, 84
Idere, 68, 90, 148
Idomaland, 120
Igala, 3, 48, 51, 52, 55, 56, 60, 69, 81, 82, 83, 84, 86, 90, 91, 98, 101, 106, 109, 110, 111, 121, 128, 129, 135, 139, 152, 181, 187, 192, 193, 203, 212, 217
Igalaland, 3, 48, 51, 52, 56, 61, 65, 69, 74, 75, 83, 84, 86, 88, 90, 91, 92, 93, 101, 103, 106, 114, 115, 121, 123, 124, 125, 129, 135, 136, 144, 147, 152, 153, 181, 187, 219

248 Visions of Sustainability

Imane, 49, 75, 76, 82, 83, 84, 85, 86, 102, 112, 119, 121, 122, 128, 129, 130, 179, 180, 181, 192, 193, 200
immigration, 88, 220
income, 5, 15, 30, 31, 36, 37, 38, 39, 41, 60, 61, 64, 68, 70, 77, 83, 89, 98, 100, 101, 104, 114, 119, 120, 127, 130, 131, 142, 149, 155, 178, 179, 180, 184, 187, 194, 195, 196, 197, 205, 206, 207, 210, 220, 232, 233, 235
indigenous institutions, 51, 105, 178, 181, 183
inflation, 48, 61, 64, 70, 72, 147, 169, 180, 183, 185, 200, 201, 206
inheritance, 56, 60, 70, 97, 140, 164, 220
integrated crop production, 7
Integrated Pest Management (IPM), 11, 20, 21, 29, 40, 144
intercrop, 12, 13, 14, 15, 28, 43, 45, 56, 64, 136, 144, 147, 158, 159, 168, 169
International Monetary Fund (IMF), 61, 200, 201
International Union for the Conservation of Nature (IUCN), 5
iron, 41, 101, 192, 212
irrigation, 11, 13, 18, 24, 35
iterative method, 53, 67

Kano, 74, 90, 117, 119, 120, 122, 128, 129, 179, 193
Kofyar, 68, 214
Kogi State, Nigeria, 2, 48, 49, 78, 85, 86, 106
kola nut, 92, 110, 136, 140, 149, 164

labour, 18, 27, 37, 50, 51, 64, 66, 68, 71, 77, 85, 93, 94, 95, 96, 99, 134, 135, 143, 145, 146, 147, 148, 149, 155, 157, 165, 167, 168, 169, 170, 174, 190, 192, 209, 220

labour group, *adakpo*, 94, 95, 148, 149, 167, 170
labour group, *ayilo*, 94, 95, 148, 167, 170
labour group, *ogwu*, 94, 95, 118, 148, 149, 167, 170
Lagos, 120, 122, 123, 126, 133, 193
Land Equivalent Ratio (LER), 15, 28, 37, 45, 136, 147, 158, 159
laterite, 83
legumes, crops, 166
legumes, trees, 137, 143, 166
leisure, 30, 51, 89, 103, 105, 178, 190, 191, 195, 206, 214, 216, 220
light, 13, 14, 146, 149, 151
lipids, 41
livelihood, 4, 14, 22, 51, 67, 77, 78, 81, 100, 103, 105, 115, 149, 155, 178, 191, 194, 195, 196, 206, 209, 210, 213, 214, 216
livelihood, diversity, 14
locust bean, 65, 85, 90, 91, 92, 94, 110, 120, 128, 129, 136, 140, 145, 149, 150, 151, 160, 164, 225, 227
low input sustainable agriculture (LISA), 7

machinery, 17, 22
madaki, 84, 85, 86, 97, 101, 106, 113, 115, 126, 184
maize, 31, 61, 89, 91, 92, 95, 109, 135, 136, 137, 139, 141, 143, 144, 145, 147, 153, 156, 157, 158, 159, 160, 163, 171, 172, 173, 174, 181, 186, 201, 211, 222, 223, 224, 230, 231
Makurdi, 74, 76
Malaysia, 29, 40
Mann-Whitney statistic, 67, 232
manure, 18, 97, 143, 166, 229
mapping, 54
Maximum Sustainable Yield (MSY), 7, 150

medicine, 89, 105, 127, 153, 178, 180, 192, 195, 213
men, 55, 58, 59, 61, 64, 65, 72, 73, 80, 84, 88, 89, 93, 94, 95, 98, 99, 100, 101, 114, 120, 124, 140, 143, 145, 146, 147, 154, 164, 178, 179, 182, 184, 185, 187, 192, 193, 194, 203, 209, 212
migration, 50, 52, 56, 68, 77, 83, 87, 92, 119, 179, 193, 199, 206
millet, 89, 90, 91, 95, 109, 120, 135, 136, 137, 139, 160, 222, 223, 224
mixtures, crop, 157

naturalistic inquiry, 53, 59
Niger, 133
Nigeria, 2, 3, 22, 23, 33, 46, 47, 48, 50, 52, 57, 58, 68, 74, 75, 78, 79, 80, 82, 83, 86, 88, 90, 96, 99, 103, 106, 114, 115, 116, 117, 118, 119, 121, 123, 124, 125, 128, 133, 135, 139, 141, 143, 148, 178, 179, 180, 183, 184, 185, 191, 192, 193, 214
nitrate, 38
nitrogen, 40, 41, 83, 136, 137, 139
nutrient recycling, 10
nutrient status, 18
nutrients, plant, 18, 83, 137
nutritional status, 36, 135

occupation, 60, 70, 72, 82, 88, 89, 100, 121, 124, 130, 178, 179, 193, 195, 197, 206, 232
ogwu, 94, 95, 118, 148, 149, 167, 170
oil palm, 55, 65, 77, 83, 92, 93, 96, 105, 110, 120, 122, 140, 145, 149, 150, 155, 160, 164, 179, 186, 192, 214, 225, 227
oja, 103, 104, 129, 181, 182, 183, 188, 190, 195, 206
okro (lady finger), 151
oral history, 54, 55, 65, 67, 84, 85, 192, 193, 212

organic agriculture, 7, 8
organic matter, 18, 19, 36, 38, 41, 216
organochlorines, 19, 20

Papua New Guinea, 18
partial LER (pLER), 45, 158, 159
participation, 2, 3, 52, 65, 82, 95
participatory rural appraisal (PRA), 52, 211
pest control, 11, 20, 23, 32, 144
pesticide, 5, 8, 10, 11, 12, 13, 19, 20, 21, 23, 24, 25, 29, 34, 35, 38, 40, 137, 144, 145, 174
pests, 6, 11, 12, 13, 19, 20, 23, 32, 36, 40, 42, 62, 71, 135, 136, 141, 143, 144, 153, 174, 209, 214, 221, 224
pigeon pea, 91, 109, 151, 230
plant nutrients, 18, 83, 137
plantain, 92, 110, 136, 140, 164
policy, 11, 24, 104
pollution, 18, 78
polygamy, 98
population, change, 77, 108
Port Harcourt, 68, 74, 127, 128, 129
potato, 31, 151
potatoes, 17
predation, 42
primary school, 88, 117, 119, 120, 121, 122, 123, 126, 127, 132, 180
productivity, 13, 16, 17, 28, 35
profit, 14, 16, 28, 30, 36, 129, 134, 135
profitability, 9, 16
protein, 151, 152, 153

quality, air, 28
quality, life, 4, 62, 114, 191, 194, 195, 205, 206
quality, system, 12
quality, water, 28
quantification, 14, 28, 29, 32, 38, 46, 114

rainfall, 37, 82, 83, 91, 149
rainwater harvesting, 149
rapid rural appraisal (RRA), 52, 211
reductionist, 32
regression analysis, 63, 66, 163, 230
remittance, 59, 70, 199
resilience, 12, 13, 14, 17, 23, 52, 141, 149, 196, 210, 217
R-factor, 141, 165
rice, 17, 42, 61, 91, 92, 109, 151, 153, 156, 171, 172, 186, 201, 224, 230, 231
River, Benue, 74, 75, 83, 181, 214, 217
River, Niger, 74, 75, 79
roofing materials, 192
rural livelihood, 6

salinization, 25, 34, 35, 38
scales, spatial, 9, 26, 27, 46
scales, time, 11, 15, 25, 26, 27, 29, 46
school, primary, 88, 117, 119, 120, 121, 122, 123, 126, 127, 132, 180
school, secondary, 117, 122, 123, 127, 132
self-reliance, 3, 31, 77, 104, 105, 191, 192, 206, 212, 213
social factors, 22, 28, 29
soil conservation, 11
soil, acrisol, 19
soil, bulk density, 37
soil, ferraisol, 19
soil, laterite, 83
soil, pH, 18, 19, 36, 83, 216
soil, quality, 18, 19, 23, 139, 141, 145, 154, 174
soil, temperature, 37
sole crop, 13, 15, 28, 43, 45, 56, 64, 92, 136, 147, 156, 157, 159, 169

soybean, 152, 153, 186
story telling, 178
strategy, coping, 52, 71, 143, 154, 166, 214, 221
strong sustainability, 6
structural adjustment programme (SAP), 3, 48, 77, 87, 137, 144, 178, 210
sugar beet, 17
Sustainability Indicators (control), 19, 20, 24, 40, 42, 56, 59, 70, 96, 101, 121, 140, 145
Sustainability Indicators (driving force)), 11, 24, 115, 155
Sustainability Indicators (pressure), 19, 22, 24, 131, 145, 214, 215
Sustainability Indicators (process), 2, 14, 24, 53, 81, 89, 94, 195
Sustainability Indicators (response), 24, 81, 138, 142, 212, 215
Sustainability Indicators (state), 11, 24, 25, 75, 80, 86, 104, 142, 181, 184, 213, 214
sustainability, strong, 6
sustainability, weak, 6
sweet potato, 31, 109, 136
system, agricultural, 6, 7, 10, 11, 17, 22, 28, 135, 216
system, agriculture, 6, 7, 10, 11, 17, 22, 28, 135, 216
system, cropping, 6, 12, 13, 14, 22, 23, 26, 27, 30, 43, 134, 135, 137, 143
system, farming, 10, 16, 30, 41, 54
system, quality, 12

teak, 93, 150
Tiv, 120, 214, 217
Total Factor Productivity (TFP), 16, 32
toxic load, 21

trade, 54, 55, 69, 72, 75, 79, 80, 89, 90, 100, 115, 117, 126, 128, 129, 130, 146, 154, 178, 179, 180, 183, 184, 189, 190, 198, 213, 217, 218
trading, long-distance, 115, 128, 179, 193, 198
trading, petty, 198
trading, short-distance, 179, 198
transect, 54
tree crop, banana, 92, 110, 136, 140, 164
tree crop, bush mango, 92, 110, 151
tree crop, cashew, 92, 110, 136, 140, 150, 164, 227
tree crop, citrus, 92, 110, 122, 140, 160, 164, 225
tree crop, cocoa, 119, 137, 151, 154, 192, 193
tree crop, coconut, 136
tree crop, coffee, 122, 137, 154, 192
tree crop, guava, 92, 110, 136, 140, 150, 164, 227
tree crop, kola nut, 92, 110, 136, 140, 149, 164
tree crop, locust bean, 65, 85, 90, 91, 92, 94, 110, 120, 128, 129, 136, 140, 145, 149, 150, 151, 160, 164, 225, 227
tree crop, oil palm, 55, 65, 77, 83, 92, 93, 96, 105, 110, 120, 122, 140, 145, 149, 150, 155, 160, 164, 179, 186, 192, 214, 225, 227
tree crop, plantain, 92, 110, 136, 140, 164
tree home gardens, 27
triangulation, 53, 54, 56, 210, 211

United Nations (UN), 24, 25, 34
USA, 12, 17, 18, 28

Vatican Council, 81
village, catchment, 46, 209, 210, 216
village, community, 26, 178
village, studies, 53

water, 24, 25, 36, 38, 39
waterlogging, 34, 35
weak sustainability, 6
weeds, 13, 36, 82, 145
Weighted Goal Programming (WGP), 30, 32, 41
wheat, 12, 17
widows, 93, 104
wildlife, 28, 37, 39
women, 55, 58, 60, 63, 64, 65, 70, 72, 73, 80, 82, 84, 88, 89, 90, 92, 93, 94, 95, 96, 97, 98, 99, 100, 101, 103, 113, 114, 115, 124, 125, 131, 135, 140, 142, 145, 146, 147, 149, 150, 151, 152, 153, 154, 164, 178, 179, 180, 183, 185, 186, 187, 189, 192, 193, 194, 203, 208, 209, 211, 212, 217
World Bank, 61, 121
World Commission for Environment and Development (WCED), 5

yam, 61, 91, 92, 96, 109, 119, 120, 136, 137, 138, 139, 140, 141, 142, 143, 144, 147, 152, 153, 157, 160, 169, 171, 174, 201, 214, 222, 223, 224, 230
yam bean, 109, 137, 138, 140, 141, 142, 143, 157, 160, 222, 223, 224
yield, actual, 44
yield, over-, 56, 147, 159
yield, theoretical, 44
Yoruba, 106, 125, 154